9–1
[BIOL]OGY
AQA [REVISION] GUIDE

Kayan Parker

Author Kayan Parker
Editorial team Haremi Ltd
Series designers emc design ltd
Typesetting Newgen KnowledgeWorks (P) Ltd, Chennai, India
Illustrations York Publishing Services and Newgen KnowledgeWorks (P) Ltd, Chennai, India
App development Hannah Barnett, Phil Crothers and Haremi Ltd

Designed using Adobe InDesign
Published by Scholastic Education, an imprint of Scholastic Ltd, Book End, Ra
Oxfordshire, OX29 0YD
Registered office: Westfield Road, Southam, Warwickshire CV47 0RA
www.scholastic.co.uk

Printed by Bell & Bain Ltd, Glasgow

1 2 3 4 5 6 7 8 9   7 8 9 0 1 2 3 4 5 6

British Library Cataloguing-in-Publication Data
A catalogue record for this book is available from the British Library.
ISBN 978-1407-17672-7

Acknowledgements

The publishers gratefully acknowledge permission to reproduce the following copyright material:

p8 Blamb/Shutterstock; p9 top photoiconix/Shutterstock; p9 bottom Lebendkulturen.de/Shutterstock; p10 Tefi/Shutterstock; p14 Zaharia Bogdan Rares/Shutterstock; p16 Pressmaster/Shutterstock; p17 Jarun Ontakrai/Shutterstock; p18 top somersault1824/Shutterstock; p18 bottom ellepigrafica/Shutterstock; p19 BlueRingMedia/Shutterstock; p20 KYTan/Shutterstock; p22 Greg Amptman/Shutterstock; p23 top Dreamy Girl/Shutterstock; p23 bottom left and right Designua/Shutterstock; p24 luchunyu/Shutterstock; p27 Jose Luis Calvo/Shutterstock; p28 Christos Georghiou/Shutterstock; p31 joshya/Shutterstock; p34 BlueRingMedia/Shutterstock; p35 Designua/Shutterstock; p36 left and middle Designua/Shutterstock; p36 right sciencepics/Shutterstock; p37 top Alila Medical Media/Shutterstock; p37 bottom toeytoey/Shutterstock; p38 BlueRingMedia/Shutterstock; p39 ellepigrafica/Shutterstock; p44 BlueRingMedia/Shutterstock; p46 Sofiaworld/Shutterstock; p49 corlaffra/Shutterstock; p53 Timonina/Shutterstock; p57 ellepigrafica/Shutterstock; p65 Den Edryshov/Shutterstock; p68 kirill_makarov/Shutterstock; p74 marina_ua/Shutterstock; p75 ducu59us/Shutterstock; p76 Praisaeng/Shutterstock; p78 Yoko Design/Shutterstock; p79 Alila Medical Media/Shutterstock; p80 top Alila Medical Media/Shutterstock; p80 bottom Designua/Shutterstock; p82 Alila Medical Media/Shutterstock; p83 Designua/Shutterstock; p84 Syda Productions/Shutterstock; p85 Sherry Yates Young/Shutterstock; p86 joshya/Shutterstock; p90 Tefi/Shutterstock; p96 top and bottom Designua/Shutterstock; p98 Sutichak/Shutterstock; p100 snapgalleria/Shutterstock; p101 Designua/Shutterstock; p103 Designua/Shutterstock; p106 left Pakhnyushchy/Shutterstock; p106 right Michiel de Wit/Shutterstock; p116 koya979/Shutterstock; p122 left Rob Hainer/Shutterstock; p122 right MusiggachartSMY/Shutterstock; p129 Patrik Dietrich/Shutterstock; p130 alinabel/Shutterstock; p131 David Lee/Shutterstock; p134 photoiconix/Shutterstock; p135 top stockshoppe/Shutterstock; p135 bottom Brendan Somerville/Shutterstock.

Every effort has been made to trace copyright holders for the works reproduced in this book, and the publishers apologise for any inadvertent omissions.

Note from the publisher:

Please use this product in conjunction with the official specification and sample assessment materials. Ask your teacher if you are unsure where to find them.

Contents

CELL BIOLOGY

Eukaryotes and prokaryotes 8
Animal and plant cells 9
Cell specialisation 10
Cell differentiation 11
Microscopy 12
Culturing microorganisms 13
Required practical 1: Using a light microscope 15
Required practical 2: Investigating the effect of antiseptics or antibiotics 17
Mitosis and the cell cycle 18
Stem cells 19
Diffusion 21
Osmosis 23
Required practical 3: Investigating the effect of a range of concentrations of salt or sugar solutions on the mass of plant tissue 25
Active transport 26
Review it! 27

TISSUES, ORGANS AND ORGAN SYSTEMS

The human digestive system 28
Enzymes 30
Required practical 4: Food tests 32
Required practical 5: The effect of pH on amylase 33
The heart 34
The lungs 35
Blood vessels 36
Blood 37
Coronary heart disease 38
Health issues 40
Effect of lifestyle on health 42
Cancer 43
Plant tissues 44
Transpiration and translocation 45
Review it! 47

INFECTION AND RESPONSE

Communicable diseases 48
Viral diseases 50
Bacterial diseases 51
Fungal and protist diseases 52
Human defence systems 53
Vaccination 54
Antibiotics and painkillers 55
New drugs 56
Monoclonal antibodies 57
Monoclonal antibody uses 58
Plant diseases 59
Plant defences 61
Review it! 62

BIOENERGETICS

Photosynthesis 63
Rate of photosynthesis 64
Required practical 6: Investigating the effect of light intensity on the rate of photosynthesis 66
Uses of glucose 67
Respiration 68
Response to exercise 70
Metabolism 71
Review it! 72

Topic 5

HOMEOSTASIS AND RESPONSE

Homeostasis	73
The human nervous system	74
Reflexes	75
Required practical 7: Investigating the effect of a factor on human reaction time	77
The brain	78
The eye	79
Focusing the light	80
Control of body temperature	81
Human endocrine system	82
Control of blood glucose concentration	83
Diabetes	84
Maintaining water and nitrogen balance in the body	86
ADH	88
Dialysis	89
Hormones in human reproduction	90
Contraception	92
Using hormones to treat infertility	93
Negative feedback	94
Plant hormones	95
Required practical 8: Investigating the effect of light or gravity on the growth of newly germinated seedlings	96
Review it!	97

Topic 6

INHERITANCE, VARIATION AND EVOLUTION

Sexual and asexual reproduction	98
Meiosis	100
DNA and the genome	101
DNA structure	102
Protein synthesis	103
Genetic inheritance	105
Punnett squares	107
Inherited disorders	109
Variation	110
Evolution	111
Selective breeding	112
Genetic engineering	113
Cloning	115
Theory of evolution	117
Speciation	118
The understanding of genetics	119
Evidence for evolution	120
Classification	122
Review it!	124

Topic 7

ECOLOGY

Communities	125
Abiotic factors	127
Biotic factors	128
Adaptations	129
Food chains	130
Measuring species	131
Required practical 9: Measuring the population size of a common species	133
The carbon cycle	134
The water cycle	135
Decomposition	136
Required practical 10: Investigating the effect of temperature in the rate of decay	137
Impact of environmental change	138
Biodiversity	139
Global warming	140
Maintaining biodiversity	141
Trophic levels	142
Pyramids of biomass	143
Food production	144
Role of biotechnology	145
Review it!	146
Glossary / Index	147
Answers	153

How to use this book

This Revision Guide has been produced to help you revise for your 9–1 GCSE in AQA Biology. Broken down into topics and subtopics it presents the information in a manageable format. Written by a subject expert to match the new specification, it revises all the content you need to know before you sit your exams.

The best way to retain information is to take an active approach to revision. Don't just read the information you need to remember – do something with it! Transforming information from one form into another and applying your knowledge through lots of practice will ensure that it really sinks in. Throughout this book you'll find lots of features that will make your revision an active, successful process.

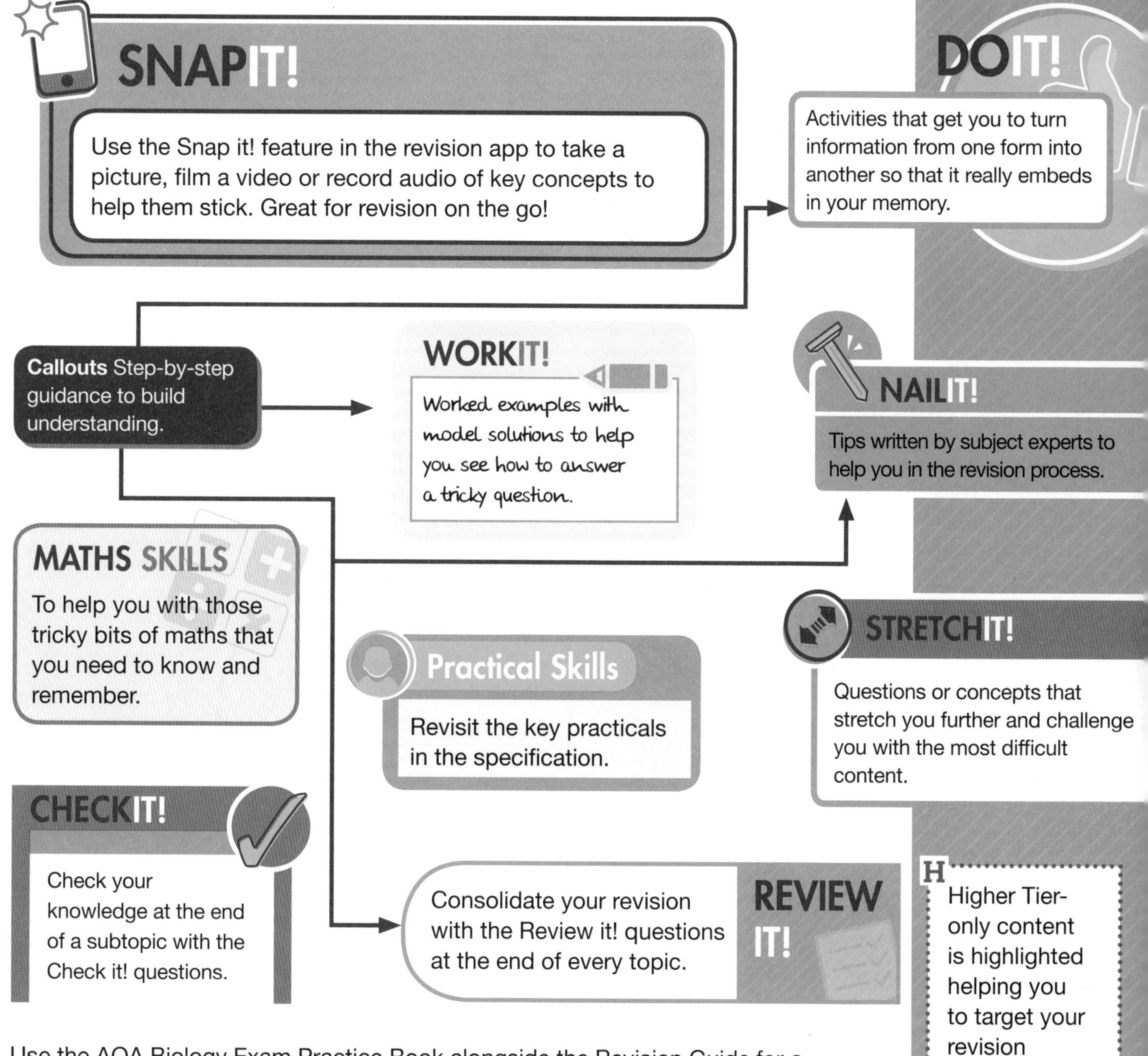

Use the AQA Biology Exam Practice Book alongside the Revision Guide for a complete revision and practice solution. Packed full of exam-style questions for each subtopic, along with full practice papers, the Exam Practice Book will get you exam ready!

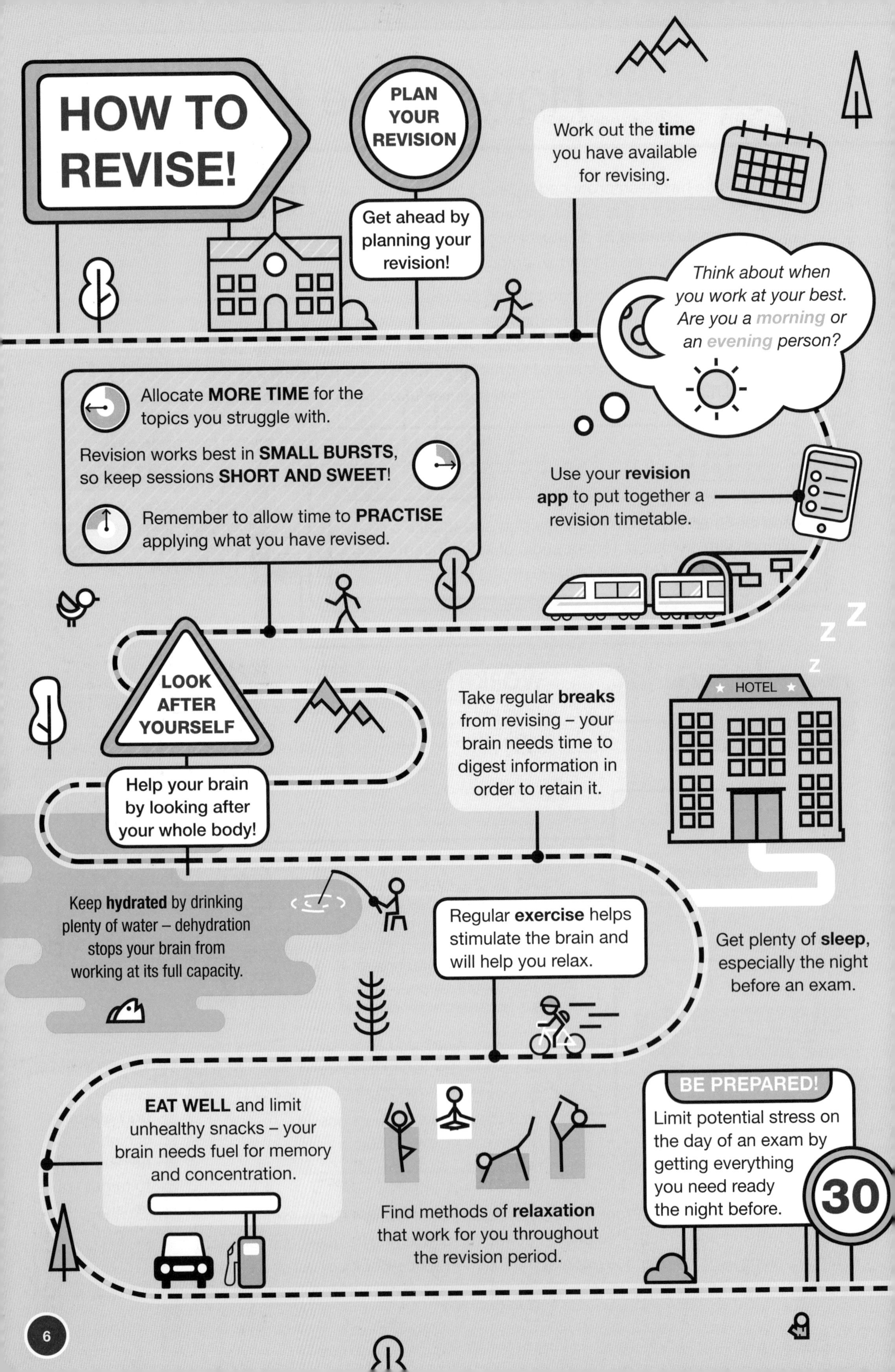
HOW TO REVISE!
PLAN YOUR REVISION
Get ahead by planning your revision!
Work out the **time** you have available for revising.
Think about when you work at your best. Are you a morning or an evening person?
Allocate **MORE TIME** for the topics you struggle with.
Revision works best in **SMALL BURSTS**, so keep sessions **SHORT AND SWEET**!
Remember to allow time to **PRACTISE** applying what you have revised.
Use your **revision app** to put together a revision timetable.
LOOK AFTER YOURSELF
Help your brain by looking after your whole body!
Take regular **breaks** from revising – your brain needs time to digest information in order to retain it.
HOTEL
Keep **hydrated** by drinking plenty of water – dehydration stops your brain from working at its full capacity.
Regular **exercise** helps stimulate the brain and will help you relax.
Get plenty of **sleep**, especially the night before an exam.
EAT WELL and limit unhealthy snacks – your brain needs fuel for memory and concentration.
Find methods of **relaxation** that work for you throughout the revision period.
BE PREPARED!
Limit potential stress on the day of an exam by getting everything you need ready the night before.
30

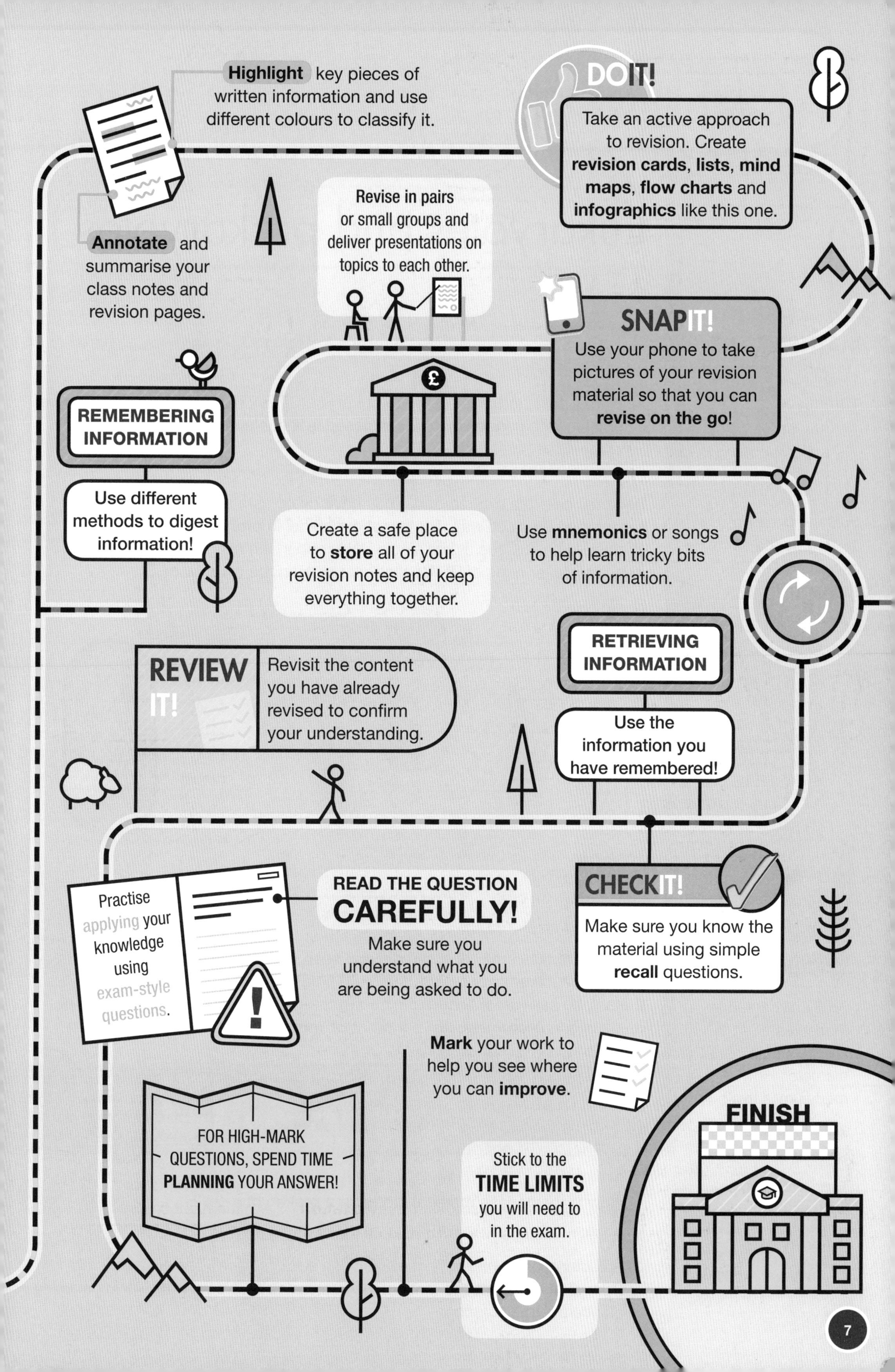
Highlight key pieces of written information and use different colours to classify it.
Annotate and summarise your class notes and revision pages.
DO IT!
Take an active approach to revision. Create revision cards, lists, mind maps, flow charts and infographics like this one.
Revise in pairs or small groups and deliver presentations on topics to each other.
SNAP IT!
Use your phone to take pictures of your revision material so that you can revise on the go!
REMEMBERING INFORMATION
Use different methods to digest information!
Create a safe place to store all of your revision notes and keep everything together.
Use mnemonics or songs to help learn tricky bits of information.
REVIEW IT!
Revisit the content you have already revised to confirm your understanding.
RETRIEVING INFORMATION
Use the information you have remembered!
Practise applying your knowledge using exam-style questions.
READ THE QUESTION CAREFULLY!
Make sure you understand what you are being asked to do.
CHECK IT!
Make sure you know the material using simple recall questions.
Mark your work to help you see where you can improve.
FOR HIGH-MARK QUESTIONS, SPEND TIME PLANNING YOUR ANSWER!
Stick to the TIME LIMITS you will need to in the exam.
FINISH

Cell biology

Eukaryotes and prokaryotes

MATHS SKILLS

1 centimetre (cm) = 10 mm

1 millimetre (mm) = 1 000 μm

1 μm = 1 000 nm

In standard form, 1×10^4 is the same as writing 10 000.

All cells are either eukaryotic or prokaryotic.

Eukaryotic cells

All animal cells and plant cells are eukaryotic. They have a cell membrane and cytoplasm with genetic material enclosed in a nucleus.

Prokaryotic cells

Bacterial cells are prokaryotic. They have cytoplasm and a cell membrane surrounded by a cell wall. The genetic material in prokaryotic cells is not enclosed in a nucleus. It is a single DNA loop in the cytoplasm. There may also be smaller rings of DNA called plasmids in the cytoplasm.

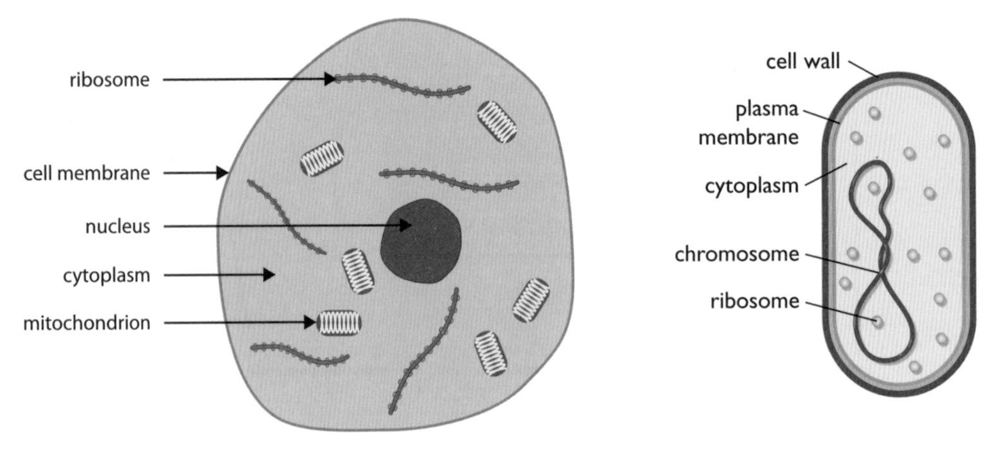

Eukaryotic cell

Prokaryotic cell

Scale and size of cells

All cells are very small and can only be seen with a microscope. Prokaryotic cells are much smaller (about one tenth smaller) than eukaryotic cells. Eukaryotic cells are measured in micrometres (μm). Prokaryotic cells can be measured in micrometers or nanometres (nm).

WORKIT!

An *E.coli* bacterium measures 2×10^3 nm in diameter. How many μm is this?

2×10^3 nm is the same as 2 000 nm.

There are 1 000 nm in 1 μm

2 000 nm / 1 000 = 2 μm

1. How is the genetic material stored in a prokaryotic cell?
2. A cell measures 5×10^3 nm. How many μm is this?
3. A bacterial cell is one-tenth the width of a eukaryotic cell. The eukaryotic cell is 2 μm wide. Calculate the width of the bacterial cell. Give your answer in nm using standard form.

Animal and plant cells

Animal and plant cells are both eukaryotic, but have some differences in their sub-cellular structure.

Animal cells

Animal cells have a nucleus, mitochondria, ribosomes, cytoplasm and a cell membrane, as well as some other sub-cellular structures.

Plant cells

Plant cells have the same sub-cellular structures as animal cells, but with the addition of chloroplasts, a permanent vacuole filled with cell sap, and a cell wall made out of cellulose. Algae also have a cellulose cell wall.

Function of sub-cellular structures

Sub-cellular structure	Function
Nucleus	Contains the genetic material
Mitochondria	Provides energy by carrying out respiration
Ribosomes	Carries out protein synthesis
Cytoplasm	Where most of the chemical reactions happen
Cell membrane	Controls the movement of substances in and out of the cell
Chloroplast	Absorbs light for photosynthesis
Permanent vacuole	Filled with cell sap to help keep the cell turgid
Cellulose cell wall	Gives strength to the cell and supports the plant
Plasmids	Additional genetic material

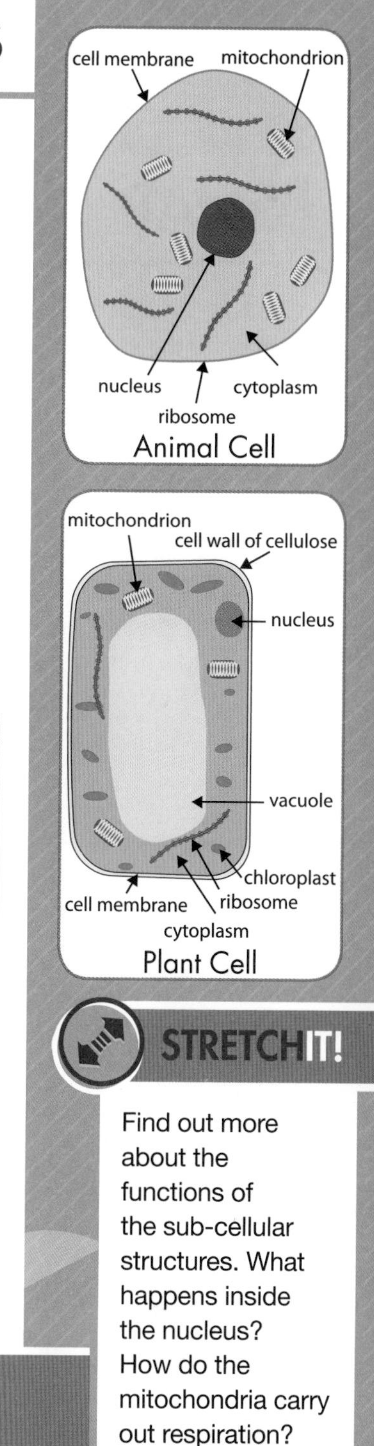

STRETCH IT!

Find out more about the functions of the sub-cellular structures. What happens inside the nucleus? How do the mitochondria carry out respiration?

CHECK IT!

1. Make a table to show which sub-structures are found in animal cells, plant cells and prokaryotic cells.
2. Explain why active cells, such as muscle cells, contain more mitochondria than less active cells.
3. This one-celled organism lives in a pond and can carry out photosynthesis. Use your knowledge of cells to identify whether this organism is a plant. Justify your answer.

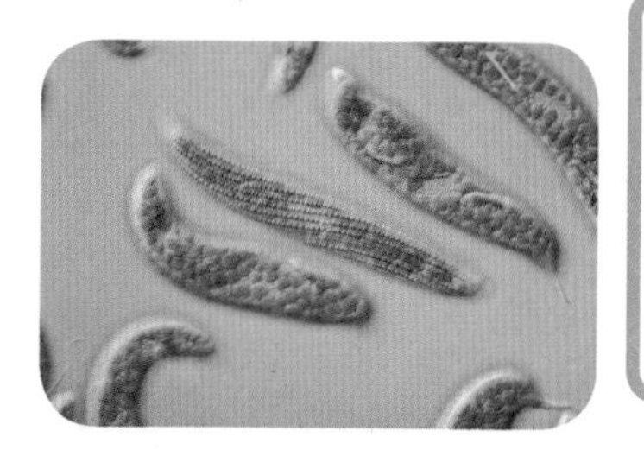

DO IT!

Look at pictures of cells in your textbook or online, and practise drawing them. Identify as many sub-cellular structures as you can.

Cell specialisation

SNAPIT!

Create your own version of this table, take a photo and learn the features of each specialised cell on the go.

Most cells are specialised in order to carry out a particular function. The cell's structure and composition are modified so that it can carry out a particular role.

Cell organisation

Cells are the basic building blocks of all organisms. Groups of similar cells come together to form a tissue, such as muscle tissue. In tissue, all the cells work together to carry out a particular function. Different tissues can come together to form organs, such as the heart. Organs are organised into organ systems, such as the circulatory system. All of the organ systems make up the whole organism.

Specialised cells

Specialised cell	Function	Specialised cell	Function
Sperm cells	Swim to the ovum (egg) for fertilisation. Have a tail for swimming. Packed full of mitochondria to provide energy. Sperm head (acrosome) contains enzymes to help break into the ovum.	Root hair cell	Take up water and mineral ions for the plant. Long thin hair to increase the surface area over which water can be taken up.
Nerve cells	Carry nerve impulses to and from the brain. Long thin axon allows nerve impulses to travel along. Many dendrites to pass nerve impulse to nearby nerve cells.	Xylem cells	Transport water from the roots to the leaves, as part of a tissue. Cells have no ends and are hollow to make a tube for water to move through. Lignin in the cell wall to waterproof the cells.
Muscle cells	Contract and relax as part of a muscle tissue, for movement. Packed full of mitochondria to provide energy.	Phloem cells	Transport sugars around the plant, as part of a tissue. Small holes in the end plates allow sugars to move through the cells.

CHECKIT!

NAILIT!

Q3 and 4 are asking you to **explain**, so give a reason for each feature, in as much detail as you can.

1. What is a specialised cell?
2. Describe how a nerve cell is specialised for carrying nerve impulses.
3. Using your knowledge of cell organisation, explain whether sperm cells can be regarded as a tissue.
4. Xylem is a specialised cell that transports water up the plant. Explain how the structure of the cell helps it to carry out this function.

Cell differentiation

In a developing embryo, none of the cells are specialised. The cells have the potential to become any type of cell. As the embryo develops, the cells begin to differentiate. They begin to become specialised to carry out a particular function.

The importance of cell differentiation

In a single-celled organism, such as a bacterium, the cell has to carry out all of the functions to keep the cell alive. In multicellular organisms, such as animals and plants, each type of cell becomes specialised to carry out one function. That way, the work is divided up by the cells. As the cells differentiate, they change their shape and acquire different sub-cellular structures.

How cells differentiate

In an embryo, the cells are not yet differentiated. These cells are called stem cells. As the cells divide, each cell is exposed to a different chemical or hormone, which makes it start to differentiate. The type of cell that the stem cell becomes depends on the hormone that it is exposed to. When a baby is born, most of the cells that make up its body are differentiated.

Stem cells

In a mature animal, cells divide only in order to replace or repair damaged tissue, and cannot differentiate into new cell types. In plants, many of the cells are able to differentiate into new cell types. This means that we are able to take cuttings from a mature plant to make many new plants.

STRETCH IT!

Find out about stem cells and how they are being used in medicine to grow new organs.

WORK IT!

Explain why differentiation is important in a multicellular organism. (3 marks)

Each type of cell can be specialised to carry out one particular function. (1)

Cells can work together to carry out functions. (1)

This division of labour is more efficient for the organism. (1)

NAIL IT!

This question is asking you to **explain**, which means that you should use your knowledge of the subject area to give a detailed account.

CHECK IT!

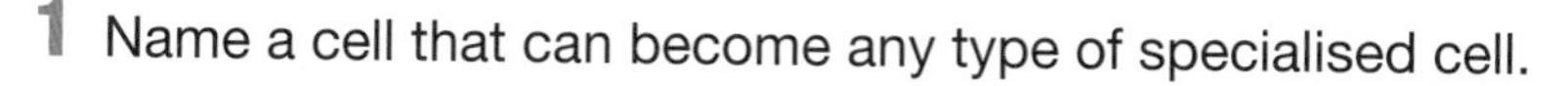

1 Name a cell that can become any type of specialised cell.

2 Where could you find cells that can differentiate into other cell types?

3 Outline how a cell becomes differentiated.

NAIL IT!

If a question asks you to **outline**, give brief steps on how to do something.

Microscopy

NAIL IT!

Magnification means how much larger the image is than the specimen.

Resolution (or **resolving power**) means how easily two points on the specimen can be distinguished from one another. The higher the resolution, the sharper the image will be.

Most cells are too small to be seen with the naked eye, so to get an understanding of what is happening inside cells we need microscopes. Microscopes have developed over the years to give higher magnifications and greater clarity.

Light microscopy

Light microscopes use light in order to view specimens. These have a low magnification and resolution (×1 500 and 200 nm), which means that the details within sub-cellular structures cannot be easily seen.

Electron microscopy

Electron microscopes use electrons to see the surface of cells, or inside the cells. These have a very high magnification and resolution (×500 000 and 0.1 nm). The sub-cellular structures within cells can be seen in detail.

MATHS SKILLS

The magnification of the cell can be worked out using the formula:

$$\text{magnification} = \frac{\text{size of image}}{\text{size of real object}}$$

This can also be shown as the magnification triangle.

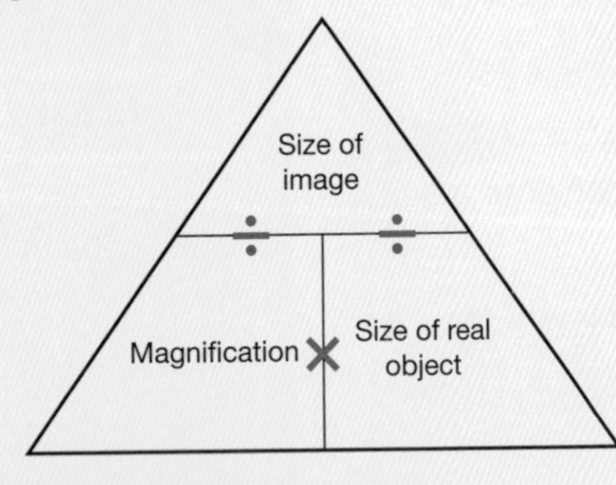

Remember 1 mm = 1 000 µm

WORK IT!

A cell that is 17 micrometers (µm) in diameter appears to be 3.4 cm in diameter when viewed through a microscope. Calculate the magnification. (3 marks)

First write the formula:

$$\text{Magnification} = \frac{\text{size of image}}{\text{size of real object}} = \frac{3.4\text{ cm}}{17\text{ µm}}$$

Then make sure that both measurements are in the same units. In this case, it will be easier to put them both into µm.

$$\text{Magnification} = \frac{34\,000\text{ µm}}{17\text{ µm}}$$

Then do the division:

Magnification = ×2 000

NAIL IT!

You should be able to write your answer in standard form.

Standard form is a way of writing very large numbers. For example:

15 000 000 is 1.5×10^7

CHECK IT!

1. Give two advantages of using an electron microscope to view cells.
2. Calculate the magnification of a cell that is 12 µm wide and appears 3 cm wide under the microscope.
3. A cell 4 µm wide was magnified 12 000 times. Rearrange the magnification formula to work out the size of the image. Write your answer in µm using standard form.

Culturing microorganisms

Microorganisms grow rapidly provided they have plenty of nutrients and oxygen, and are at the optimum temperature and pH. It is important that the microorganism in your culture is the one you want, so aseptic techniques are used to keep out other microorganisms.

Bacterial reproduction

Bacteria reproduce by a process called binary fission. This is a form of simple division, where the bacterium doubles in size and then divides into two daughter cells. Some bacteria can divide in as little as 20 minutes. Bacteria can be grown in a nutrient broth (or culture media) or on an agar plate. These both contain all of the nutrients that the bacteria need to live.

Aseptic techniques

Bacteria are used to test the effectiveness of disinfectants and antibiotics. Therefore it is important that bacteria are not contaminated with other microorganisms. To prevent contamination, aseptic techniques are used. These include:

- sterilising all Petri dishes and culture media
- sterilising inoculation loops by passing them through a flame
- securing the lid of the Petri dish with tape and storing it upside down
- not incubating bacterial cultures above 25°C.

NAIL IT!

Growing bacterial cultures below 25°C will slow down their rate of reproduction. This is safer to use in schools and colleges.

Practical Skills

Preparing an uncontaminated culture

1. Wear a lab coat and gloves.
2. Take a sterilised Petri dish or conical flask containing culture media.
3. Pass an inoculation loop through a Bunsen flame, cool, and then dip it into culture media containing your desired bacteria.
4. Using the inoculation loop, spread the bacterial sample over the surface of the Petri dish, or place inside the conical flask and stir. Quickly replace the lid.
5. Pass the inoculation loop through the flame again to sterilise it.
6. Secure the lid of the Petri dish or conical flask with tape. Place the Petri dish upside down.
7. Leave the bacterial culture to grow at a maximum temperature of 25°C.

DO IT!

Make cue cards of these seven steps and jumble them up. Practise putting the cards in the correct order.

MATHS SKILLS

Calculating cross-sectional area

You may be asked to calculate the cross-sectional area of a bacterial colony or the clear area around a colony. For this, you should use the formula, πr^2.

WORKIT!

The clear area around a bacterial colony is 11 mm in diameter. Calculate the cross-sectional area of the clear area. (3 marks)

Cross-sectional area $= \pi r^2$

Cross-sectional area $= 3.142 \times 5.5^2$

Cross-sectional area $= 3.142 \times 30.25$

Cross-sectional area $= 95\,mm^2$

NAILIT!

You will be expected to write your answer in standard form.

MATHS SKILLS

Calculating number of bacteria in a population

You should be able to calculate the number of bacteria in a population after a certain time so long as you know the starting number and the mean division time.

WORKIT!

A bacterial colony of 100 000 bacteria divides every 24 minutes. How many bacteria will there be after six hours? (3 marks)

First work out how many divisions there have been.

Six hours is 360 minutes.

$360 \div 24 = 15$ divisions.

Now work out how many bacteria there will be after 15 divisions.

$100\,000 \times 2^{15} = 3\,276\,800\,000$ bacteria

3.3×10^9 bacteria

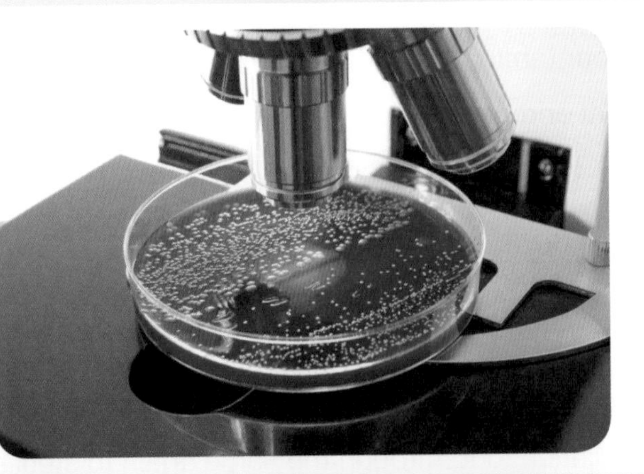

CHECKIT!

1 Describe the process of bacterial reproduction.

2 Name two aseptic techniques.

3 Calculate the cross-sectional area of a bacterial colony that is 400 μm in diameter. Use the formula, πr^2.

Required practical 1: Using a light microscope

This is the first of ten practicals you are required to carry out for GCSE biology and that you could be questioned on in the exam.

For this particular practical you need to produce labelled scientific drawings using a light microscope. A magnification scale must be used. Make sure you know how to calculate the magnification of a specimen observed through a light microscope.

Observing plant and animal cells

How to set up a light microscope

1. Place the specimen on the stage.
2. Switch on the microscope so that the light passes through the specimen.
3. Make sure that the ×4 objective lens is clicked into place above the specimen.
4. Bring the specimen into focus by looking down the eyepiece lens and moving the coarse focus.
5. When the specimen is in focus, move the objective lenses so that the ×10 objective lens is clicked into place above the specimen.
6. If the specimen is out of focus, bring it into focus using small movements of the fine focus.
7. Repeat steps 5 and 6 with the ×40 objective lens.
8. You should now be able to observe your specimen.

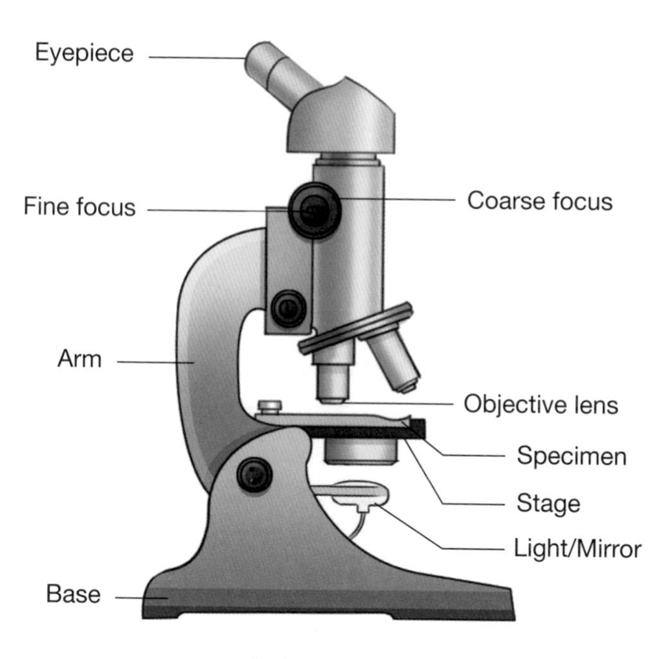

A light microscope

Practical Skills

The magnification of the specimen will be the eyepiece lens multiplied by the objective lens.

In this case, it will be ×4 multiplied by ×10 = ×40 magnification.

Practical Skills

- Always use a sharp pencil.
- Draw a smooth line (no shading or sketching).
- Make drawings as big as the space allows.
- Label lines should be drawn with a ruler.
- Include a magnification scale.

MATHS SKILLS

Learn the magnification triangle below by heart.

You may be asked to rearrange the formula in the exam.

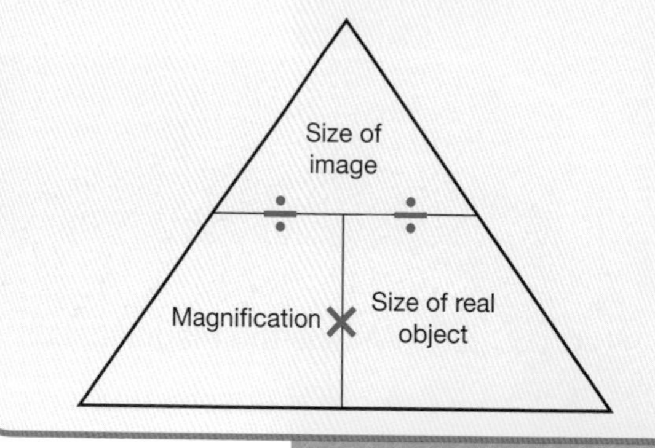

Drawing and labelling plant and animal cells

You will need to be able to draw and label the plant and animal cells that you see.

WORKIT!

Assume that this animal cell is 1 mm in width as seen through the microscope, at a magnification of ×100. Work out the actual size of the cell.

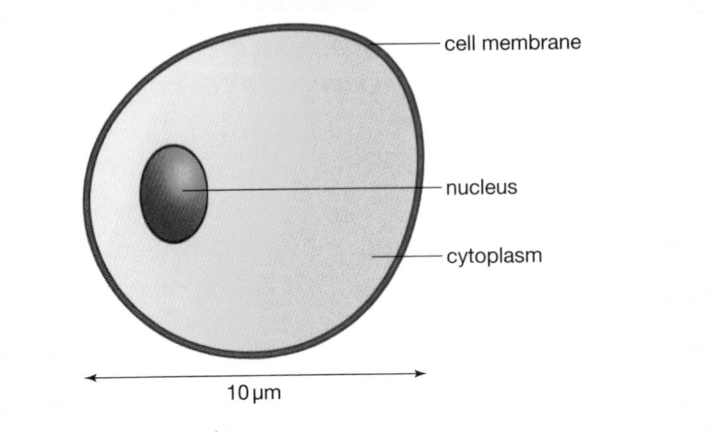

1 mm divided by 100 = 0.01 mm or 10 µm. The width of the cell is 10 µm.

DOIT!

Practise drawing and labelling plant and animal cells by drawing photos of cells found in your textbook or online. Try to identify the nucleus, cytoplasm, cell membrane and any other organelles you can see.

CHECKIT!

1 What is the magnification if the eyepiece lens is ×10 and the objective lens is ×40?

2 A cell is observed using a microscope at a magnification of ×400. The cell image is 2 mm in diameter. Calculate the actual size of the cell.

3 Outline how a student could observe some blood cells using a light microscope.

Required practical 2: Investigating the effect of antiseptics or antibiotics

For this practical, you may be asked to plan experiments to test your own hypotheses about the effectiveness of different antibiotics or antiseptics. You will need to record accurate measurements and observe standard health and safety precautions when using bacterial cultures.

Practical Skills

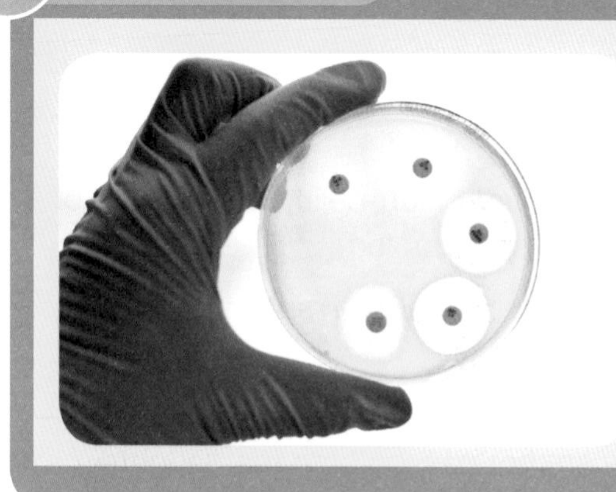

Clear areas around your bacteria colonies on the agar plates show if the **antibiotic** or **antiseptic** inhibits bacterial growth. These clear areas are called **zones of inhibition**.

Practical Skills

You will need to select appropriate apparatus and techniques when planning this investigation.

For more on aseptic techniques, see page 13.

WORKIT!

Plan an investigation to test the effect of three antibiotics on the growth of bacteria. (4 marks)

Use aseptic techniques to spread bacteria onto an agar plate.

Grow at 25°C for several days, until bacterial colonies can be clearly seen.

Place a drop of each antibiotic solution onto different areas of the agar plate and leave for 24 hours.

Observe the agar plate and measure any zones of inhibition.

MATHS SKILLS

You may be asked to work out the cross-sectional area of the zones of inhibition, or the bacterial colonies. You need to use the formula, $A = \pi r^2$ (see page 14).

CHECKIT!

1. What do clear areas around a bacteria colony show?
2. Explain why it is important to use aseptic techniques when growing bacteria.
3. A bacteria colony has a radius of 0.5 cm. Calculate the cross-sectional area A of the colony. Use the formula $A = \pi r^2$

Mitosis and the cell cycle

WORKIT!

Why is it important that the chromosomes are checked during the G2 phase? (2 marks)

The chromosomes have to be checked for errors. (1)

If the errors are passed onto the daughter cells, it could lead to mutations. (1)

Most of the cells in the body go through the cell cycle. The final stage of the cell cycle is mitosis, a type of cell division.

Chromosomes

The nucleus of a cell contains chromosomes made out of DNA. Each chromosome contains many genes. In the body cells, the chromosomes are found in pairs. In human body cells, there are 23 pairs of chromosomes.

The cell cycle

The cell cycle is divided into four stages:

G1 phase – First gap phase in which the sub-cellular structures of the cell (except the chromosomes) are doubled.

S phase – During the synthesis phase, the DNA replicates to form two copies of each chromosome.

G2 phase – The second gap phase. Here, the chromosomes are checked for errors, so that they are not passed onto the daughter cells.

M phase – In the mitosis phase, the cell divides into two identical daughter cells.

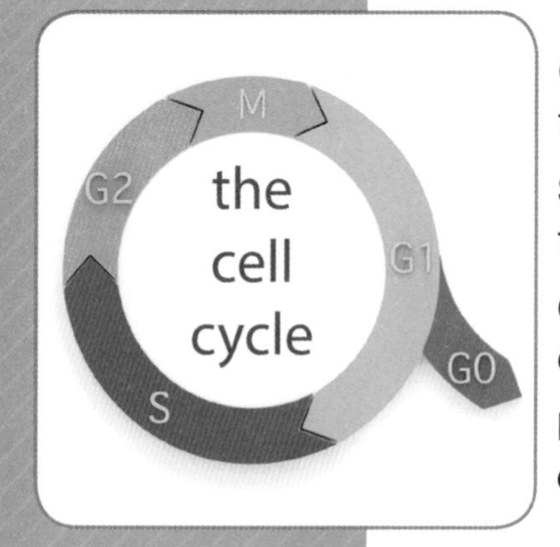

The cell cycle

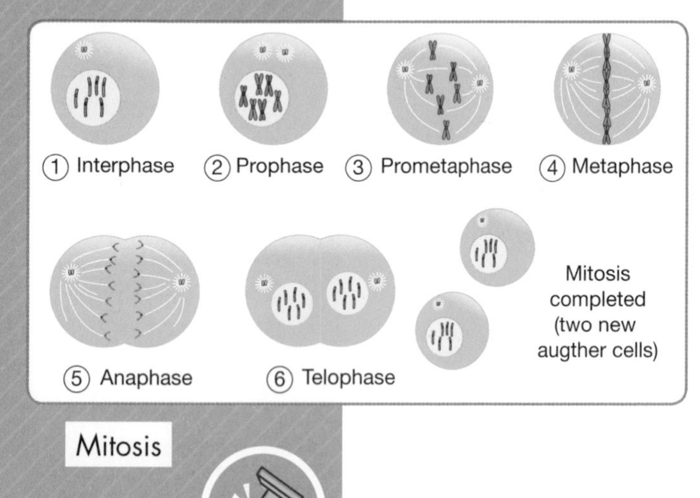

Mitosis

Mitosis

During mitosis, the doubled chromosomes line up in the centre of the cell. One set of chromosomes is pulled to each end of the cell, and the nucleus divides to form two new nuclei. The cytoplasm and cell membranes then divide to form two identical cells.

Why does mitosis occur?

Mitosis is the process for:

- growth of multicellular organisms
- repair of damaged tissues
- replacement of cells
- asexual reproduction (single-celled organisms and some plants).

NAILIT!

You do not need to know the name of each stage of mitosis, but you do need to know what is happening to the chromosomes and the nucleus.

CHECKIT!

1 Give three reasons for mitosis in a plant.

2 Describe what happens during mitosis.

3 Calculate how many cells there would be if a single cell divided by mitosis 24 times.

Stem cells

Stem cells are undifferentiated cells that can become any type of specialised cell.

Stem cells in plants and animals

Stem cells are found in developing embryos. As the cells divide, they are exposed to different chemicals and hormones and begin to differentiate into specialised cells (see page 10). Some stem cells can be found in adults, but these have a limited number of specialised cells that they can become. Plants contain stem cells called meristem tissue in the rapidly dividing root and shoot tips.

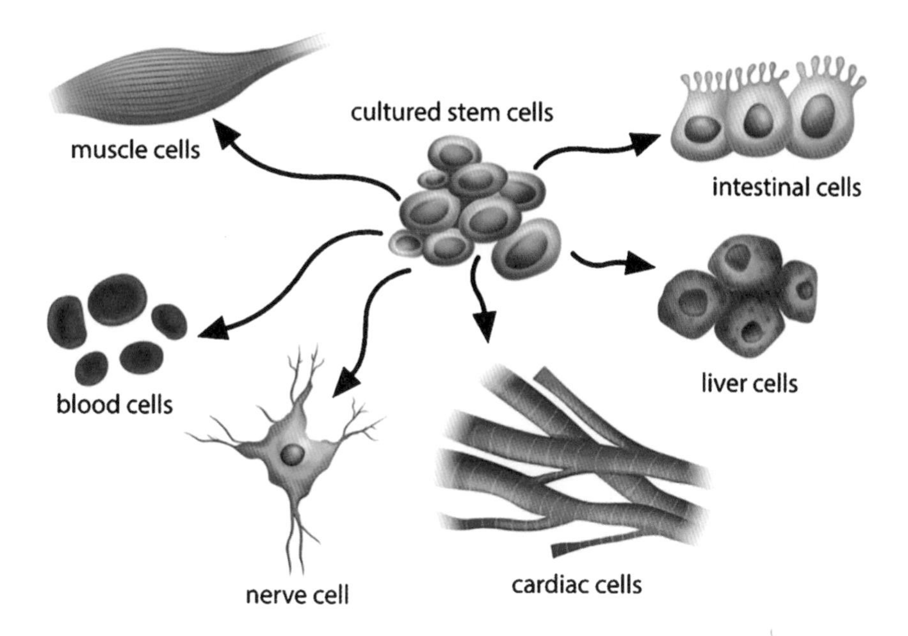

Functions of stem cells

- Stem cells in embryos become all the specialised cells in the body.
- Adult stem cells are found in different organs around the body. They replace cells in that particular organ. For example, stem cells in the skin can differentiate into different types of skin cell.
- Stem cells from adult bone marrow can form many types of cell including blood cells.
- The plant root tips and shoot tips contain meristem tissue that can become specialised cells.

Therapeutic cloning

Stem cells can be used to grow organs for transplant. An embryo is produced that has the same genes as the patient. Stem cells from this embryo can be used to grow an organ that will not be rejected from the patient's body. Stem cells can also be injected for other medical treatments. For example, treatment with stem cells may be able to help conditions such as diabetes and paralysis.

DO IT!

Make revision cards with the functions of stem cells/meristem tissue on one side and how this links to one of their uses on the other side.

Advantages	Disadvantages
No rejection of cells/organs by patient	Transfer of viral infection
No waiting time for transplants	Ethical/religious objections

Uses of stem cells in plants

- Cuttings taken from the meristem tissue in plants can be used to produce clones of plants quickly and economically.
- Meristem tissue is found in plants where new growth takes place, for example, root and shoot tips and flowering parts.
- Rare species of plants can be cloned to protect them from extinction.
- Crop plants with special features, such as disease resistance, can be cloned to produce large numbers of identical plants for farmers.

STRETCHIT!

Find out how cuttings from plants can be used to produce many cloned plants.

WORKIT!

Evaluate the practical risks and benefits, as well as social and ethical issues, of the use of stem cells in medical research and treatments. (4 marks)

Stem cells from embryos can be used to research many diseases that affect humans. (1) However, there is an ethical objection to using embryos as they could potentially grow into humans/animals. (1) Using stem cells in medical treatments means that the body will not reject the cells, (1) but there is a risk of transfer of viral infection from putting the stem cells into the body. (1)

NAILIT!

This question is asking you to **evaluate**, which means that you should make a judgement about the use of stem cells using the available evidence.

CHECKIT!

1 Where is meristem tissue found in a plant?

2 Give two uses of stem cells.

3 Biologists are trying to conserve a rare species of orchid. Describe how they could use the properties of meristem tissue to do this.

Diffusion

Substances may move in and out of cells by diffusion.

What is diffusion?

Diffusion is the spreading out of the particles of any substance in solution, or particles of a gas, resulting in a net movement from an area of higher concentration to an area of lower concentration. In cells, diffusion happens across the cell membranes, to allow substances such as oxygen and carbon dioxide to get in and out of the cell. Diffusion also allows the waste product, urea, to get out of the cell.

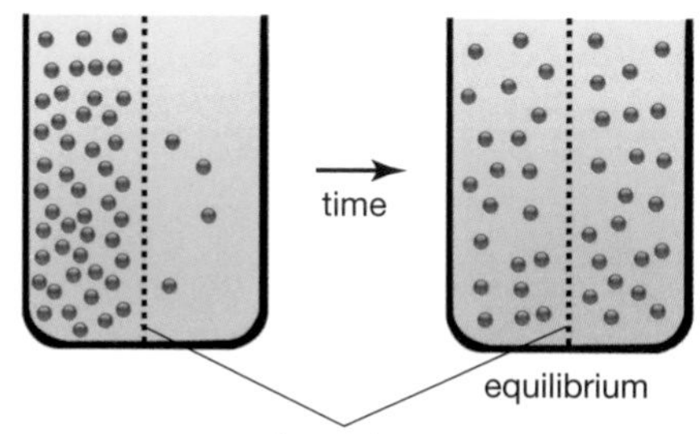

The diagram above shows the movement of particles across the cell membrane over time. The particles move from the more concentrated side to the less concentrated side, until both sides have the same number of particles (equilibrium).

Factors that affect the rate of diffusion

The rate of diffusion will increase if:

- The difference in the concentrations (the concentration gradient) increases.
- The temperature increases. Increasing the temperature increases the kinetic energy of the particles, which makes them move faster.
- The surface area of the membrane increases. Areas of the body with a large surface area, such the alveoli of the lungs, have a faster rate of diffusion.

Here are some other examples of working out the surface area to volume ratio:

Cube side (cm)	Surface area (cm^2)	Volume (cm^3)	Surface area to volume ratio
1	6	1	6:1
3	54	27	2:1

SNAP IT!

Draw out the diagrams on this page, page 23 and page 26. Compare them to learn the differences between diffusion, osmosis and active transport. Take a photo to help with your revision.

MATHS SKILLS

Surface area to volume ratio is best understood if you imagine the organisms to be shaped into cubes.

WORK IT!

What is the surface area to volume ratio of a cube that has sides that are 2 cm in length? (3 marks)

The surface area of each side is $2\,cm \times 2\,cm = 4\,cm^2$.

However, the cube has six sides, so the surface area is $4\,cm^2 \times 6 = 24\,cm^2$.

The volume is $2\,cm \times 2\,cm \times 2\,cm = 8\,cm^3$.

The surface area to volume ratio is 24:8.

This can be simplified to 3:1.

The need for exchange surfaces and transport systems

The smaller the cube (organism), the larger the surface area to volume ratio. For example, single-cell organisms, such as bacteria, have a relatively large surface area, compared to their volume. This allows sufficient transport of particles into and out of the cell to meet their needs.

Large, multicellular animals have much smaller surface area to volume ratios. This means that they cannot get all of the substances they need by diffusion alone. They need specialised exchange surfaces in order to get all of the substances that they need.

The effectiveness of the specialised exchange surface is increased when:

- There is a large surface area
- The membrane is thin, to provide a short diffusion pathway
- There is a good blood supply (in animals) to remove the diffused particles, and help to maintain a high concentration gradient
- It is ventilated (in animals for gas exchange).

Specialised exchange surfaces

Mammalian lungs – many small alveoli give the lungs a large surface area for the diffusion of oxygen and carbon dioxide. A good blood supply helps to maintain a large concentration gradient.

STRETCHIT!

Look at a diagram of villi and microvilli to understand how the surface area of mammalian intestines is increased.

Mammalian small intestines – villi and microvilli give the inner surface of the small intestine a large surface area for the diffusion of the products of digestion. A good blood supply helps to maintain a large concentration gradient.

Fish gills – have a large surface area for the diffusion of oxygen and carbon dioxide.

Plant roots – root hair cells have long hairs that increase the surface area for the movement of water (by osmosis – see page 23).

Plant leaves – have large air spaces inside the leaf and many holes (stomata) on the underside of the leaf, for the diffusion of carbon dioxide and oxygen.

You can see the gills clearly on this shark

CHECKIT!

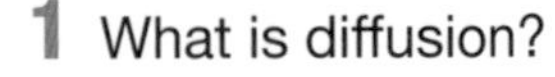

1 What is diffusion?

2 Name two ways in which the rate of diffusion can be increased.

3 A cube has sides 4 cm in length. What is the surface area to volume ratio?

Osmosis

Water may move across cell membranes by osmosis.

What is osmosis?

Osmosis is the movement of water from a dilute solution to a concentrated solution through a partially permeable membrane.

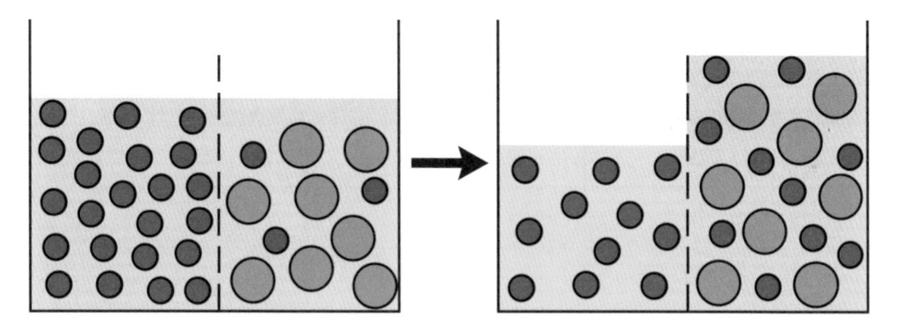

If cells are in a dilute solution, water moves into the cells. This causes plant cells to swell and become turgid, and animal cells to swell and eventually burst. If cells are in a concentrated solution, water moves out of the cell. This causes plant cells to become flaccid and plasmolysed, and animal cells to shrivel and crenate.

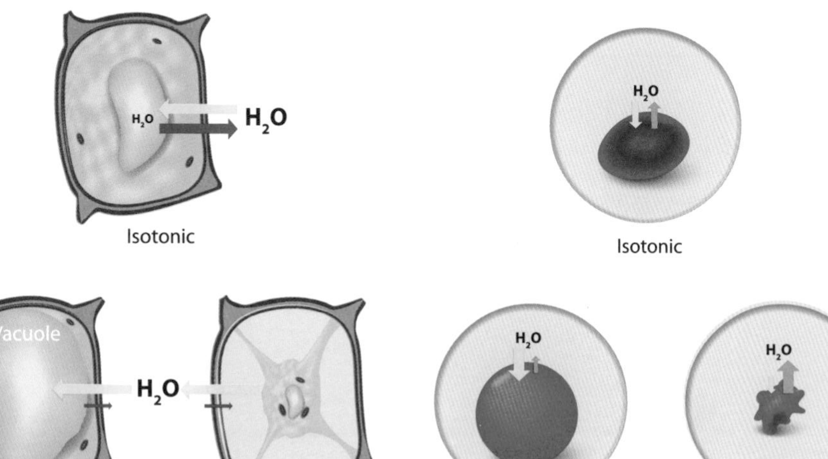

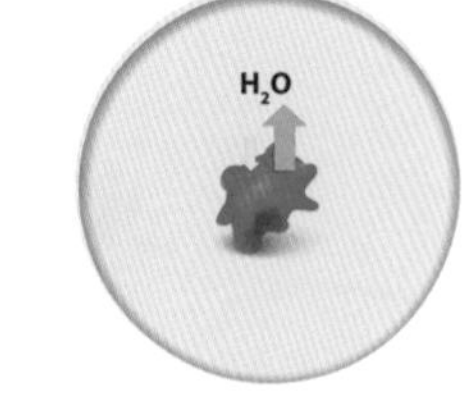

H_2O

Hypertonic

SNAP IT!

Take a photo of the tonicity diagram below and learn the 3 types of tonicity. Make sure you can describe them.

Measuring the rate of water uptake

The amount of water taken up by osmosis can be measured using a Visking osmometer. A piece of Visking tubing containing a concentrated solution, is placed into a beaker of water. A glass tube is placed into the top of the Visking tubing. As the water from the concentrated solution moves into the Visking tubing, the liquid level in the glass tubing rises. This allows the volume of water that moved into the Visking tubing to be measured.

Measuring the change of mass in plants

It is possible to measure the rate of osmosis by placing plant tissue in solutions of different concentrations, and measuring the mass before and after using a weighing balance. If the plant tissue is placed into a dilute solution, water moves into the tissue, so the mass increases. If the plant tissue is placed into a concentrated solution, water moves out of the tissue, and the mass decreases.

WORKIT!

Calculate the percentage change in mass of plant tissue placed into 0%, 5% and 10% salt solutions. (3 marks)

Concentration (%)	Mass before (g)	Mass after 4 hours (g)	% change in mass
0	10	15	+50
5	10	12	+20
10	10	8	-20

$$\text{Percentage change} = \frac{15-10}{10} \times 100\%$$
$$= \frac{5}{10} \times 100\%$$
$$= 0.5 \times 100\%$$
$$= 50\%$$

If it is a percentage increase, add a plus sign, if it is a percentage decrease, add a minus sign.

To work out the percentage change, subtract the mass before from the mass after and divide by the mass before. Then multiply by 100%.

NAILIT!

When plotting a graph, make sure the independent variable (in this case, the salt solution concentration) is on the *x*-axis, and the dependent variable (in this case, the percentage change in mass) is on the *y*-axis.

CHECKIT!

1. What will happen to an animal cell if it is placed into a concentrated solution?
2. A disc of plant tissue with a mass of 8 g is placed into a dilute solution. After four hours, the mass of the disc has increased to 12 g. Calculate the percentage increase in mass.
3. Using the data from the table in the worked example, draw a graph to show the relationship between percentage change in mass and concentration.

Required practical 3: Investigating the effect of a range of concentrations of salt or sugar solutions on the mass of plant tissue

In this practical you will use your knowledge of osmosis (see page 23) to create hypotheses about plant tissue and plan experiments to test these hypotheses. You will need to recall how to work out the percentage change in mass.

Practical Skills

To investigate the effect of different concentrations of sugar/salt solutions on the mass of plant tissue (usually potato). You should:

- Measure the mass in grams before and after the tissues had been in the solutions.
- Calculate the percentage change in mass.

WORKIT!

A cube of potato had a mass of 5.0 g, which decreased to 3.8 g after being in sugar solution for 4 hours. What conclusions can you make about the sugar solution? (3 marks)

Water has moved out of the potato by osmosis (1) from a dilute solution to a concentrated solution. (1) The sugar solution is a more concentrated solution than the contents of the potato cells. (1)

NAILIT!

When measuring the effect of concentration, other variables must be kept constant to make it a fair test. These are the **control variables.**

NAILIT!

If the mass has decreased, then water has moved out of the tissue.

MATHS SKILLS

You will be asked to work out the percentage change in mass for each piece of plant tissue. You should draw a graph of the percentage change in mass.

CHECKIT!

1 The graph shows the percentage change in mass of potato cubes placed into sugar solutions of 0, 1, 2, 3, and 4% concentration for 4 hours. Each potato cube was initially 5.0 g in mass.

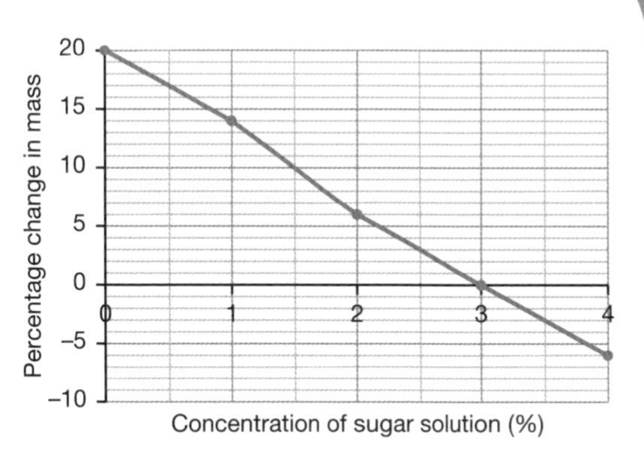

a Calculate the percentage change in mass at 2% concentration.

b Calculate the mass of the potato cube at 2% concentration after it had been in the sugar solution for four hours.

c State two variables that must be kept constant during this investigation.

Active transport

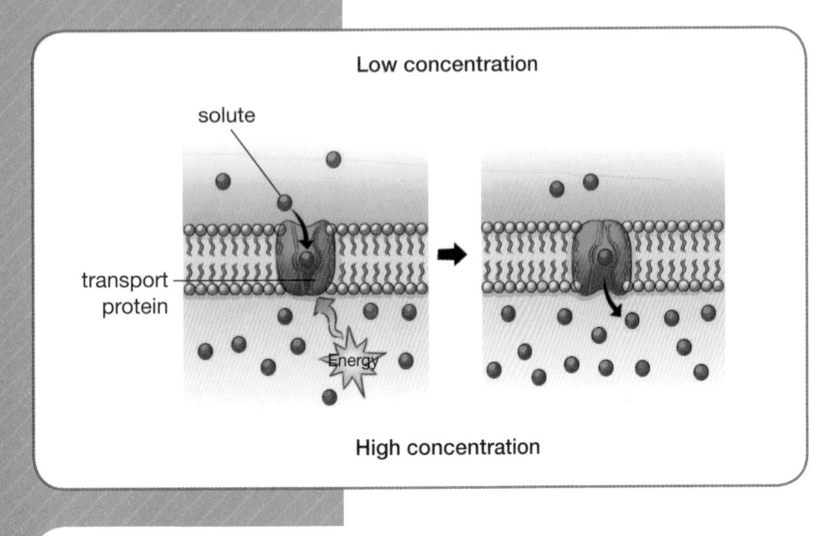

Active transport is the movement of particles against the concentration gradient.

Active transport needs energy

Sometimes, cells need to take in, or release substances against the concentration gradient. This requires energy from respiration. The substances are transported from a more dilute solution to a more concentrated solution across the cell membrane using carrier proteins.

Active transport in plants

Root hair cells in plants need to take up mineral ions from the soil for healthy growth. There is already a high concentration of minerals ions inside the cell, so mineral ions need to move from a dilute solution in the soil to a concentrated solution inside the cell. The root hair cell uses energy to transport the mineral ions through protein carriers in the cell membrane.

Active transport in humans

In our bodies, sugar molecules need to be transported from relatively low concentrations in the small intestine to higher concentrations in the blood. Sugar is needed for respiration, so sugar molecules are actively transported across the wall of the small intestine into the blood.

NAILIT!

This question is asking you to **explain**, so you should compare the properties of each type of transport.

WORKIT!

Explain the differences between diffusion, osmosis and active transport.

Many substances can move by diffusion or active transport, but only water can move by osmosis. (1)

Diffusion is the movement of particles from a concentrated solution to a dilute concentration, but active transport and osmosis involve the movement of particles or water from a dilute concentration to a concentrated solution. (1)

Diffusion and osmosis do not require energy, but active transport does require energy. (1)

CHECKIT!

1 What is meant by the concentration gradient?

2 Where does the energy for active transport come from?

3 Describe how mineral ions are taken into the plant root hair cell.

Cell biology

REVIEW IT!

1 A plant cell is a eukaryotic cell. List three ways in which a prokaryotic cell is different to a eukaryotic cell.

2 a Describe how to use a light microscope to observe a specimen.

b A magnified image of a cell is 30 000 µm in diameter and the actual diameter of the cell is 10µm. What is the magnification?

3 a Describe how a root hair cell is specialised and explain how these adaptations help the cell to carryout its function.

b How does the root hair cell become specialised?

4 Some students investigated the growth of bacterial colonies on a Petri dish that had been treated with four different antibiotics A – D.

a Plan an investigation to do this.

b What are the independent and dependent variables in this investigation?

The results are shown below:

Antibiotic	Radius of colony (cm)	Cross-sectional area of colony (cm^2)
None	2.2	15.2
A	1.8	
B	0.2	
C	0.8	2.0

c Calculate the cross-sectional area of the colonies treated with antibiotics A and B.

d Use evidence from the table to explain which antibiotic was the most effective at inhibiting bacterial growth.

e Explain why the students used aseptic techniques in their investigation.

f How could the students have made sure that their investigation was valid and reliable?

5 a What is the correct order of the stages of mitosis?

_____________ metaphase _____________ _____________

The photograph below shows onion root tip cells in the process of mitosis.

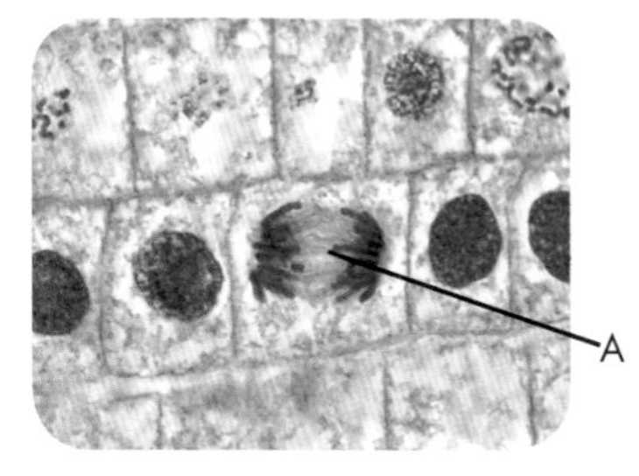

b What stage of the cell cycle is cell A in?

c Explain why onion root tip cells are undergoing mitosis.

6 Stem cells are used in research to make organs for transplant.

a What are stem cells?

b Describe where stem cells can be found.

c Discuss the advantages and disadvantages of using stem cells to make organs for transplant.

7 a Define diffusion.

b The diagram below shows the concentration of salt ions on either side of a partially permeable membrane. In which direction will the salt ions move?

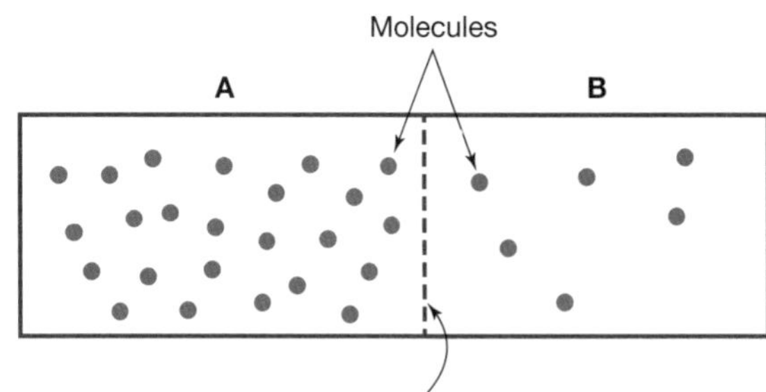

c Explain your answer to (b).

8 a What type of diffusion requires protein channels in order to cross the cell membrane?

b Compare diffusion and active transport.

The human digestive system

The human digestive system breaks down all of the food that you eat; mechanically, by chewing, and chemically, using enzymes.

Which organs are parts of the human digestive system?

The human digestive system is made up of the stomach, small intestine, large intestine, pancreas, liver and gall bladder.

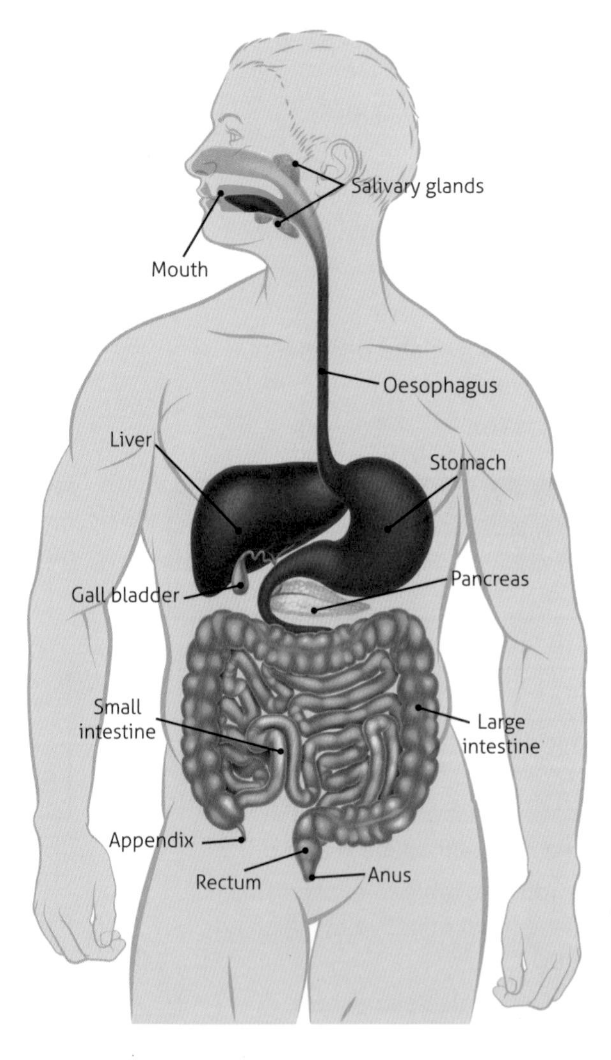

SNAPIT!

Find a picture of the digestive system, take a photo and learn it. You may be asked to label a similar diagram in the exam.

Chewed food mixes with saliva from the salivary glands, and then travels down the oesophagus to the stomach. Large lumps of food are broken down into smaller lumps in the stomach, before moving into the small intestine. Bile from the liver, and pancreatic juices from the pancreas are added to the food in the small intestine. These help with digestion.

Food needs to broken down into small, soluble molecules so it can be absorbed into the bloodstream. Any food that is not digested (such as fibre) moves into the large intestine. Here, water is removed and a solid mass known as faeces passes out of the body through the anus.

Enzymes used in digestion

Enzymes	Site of production	Substrate	Products	Uses
Carbohydrase (Amylase)	Salivary glands (carried in saliva) Pancreas (carried in pancreatic juices)	Carbohydrate (starch)	Sugars	Respiration Builds new carbohydrates
Proteases	Stomach Pancreas (carried in pancreatic juices)	Protein	Amino acids	Builds new proteins
Lipases	Pancreas (carried in pancreatic juices)	Lipids (fats)	Fatty acids Glycerol	Builds new lipids

Bile is not an enzyme, but it is made by the liver and secreted by the gall bladder, into the small intestine, to help with lipid digestion. Bile does not digest lipids, but breaks them into smaller droplets (emulsifies) so that the lipase enzymes have a greater surface area to work on. Bile is alkaline, so it also neutralises hydrochloric acid from the stomach. This is important because the enzymes in the small intestine work best at a slightly alkaline pH.

Products of digestion

The products of digestion can be written as word equations:

$$\text{starch} \xrightarrow{\text{amylase}} \text{sugars}$$

$$\text{protein} \xrightarrow{\text{protease}} \text{amino acids}$$

$$\text{fat} \xrightarrow{\text{lipase}} \text{fatty acids} + \text{glycerol}$$

DOIT!

Learn these word equations by heart. You may be asked to write them out in an exam.

WORKIT!

Explain how new proteins are made in the body. (3 marks)

Proteases are made in the stomach/pancreas. (1)

Proteases break down protein from foods into amino acids. (1)

Amino acids are combined to make new proteins in the body. (1)

1 Where in the body are lipases made?

2 What is the role of amylase?

3 Explain why bile is needed in the digestion of lipids.

Enzymes

Enzymes increase the rate of chemical reactions in living organisms.

What is an enzyme?

An enzyme is a biological catalyst. Enzymes allow chemical reactions to take place faster at lower temperatures than non-biological catalysts. All of the metabolic reactions in the body (such as breaking down or synthesising molecules) require enzymes.

The effect of temperature and pH on enzymes

Enzymes have specific (optimum) temperatures and pH at which they work best. Enzymes in the human body have an optimum temperature of 37°C. The optimum pH of the enzyme depends on where it acts.

- Amylase in saliva – pH7
- Pepsin in the stomach – pH2
- Lipase in the small intestine – pH8

The effects of temperature and pH on enzymes are shown in the table:

	Below optimum	Above optimum
Temperature	Less kinetic energy, so enzymes and their substrate move more slowly and fewer collisions occur. The rate of reaction decreases.	High temperatures denature the enzyme by breaking hydrogen bonds in its structure. The enzyme's active site changes shape and is no longer able to function.
pH	Loss of activity for the enzyme. The enzyme's active site changes shape and the rate of reaction decreases.	Loss of activity for the enzyme. The enzyme's active site changes shape and the rate of reaction decreases.

MATHS SKILLS

Carrying out rate calculations

The rate of reaction is the speed at which a reaction is taking place. To work out the rate of reaction, divide the amount of reactant used, or product made, by time.

WORKIT!

Hydrogen peroxide is converted into oxygen and water by the enzyme catalase. The speed of the reaction can be worked out by measuring the amount of oxygen collected in 5 minutes. Calculate the rate of reaction if 15 cm^3 of oxygen is collected in 5 minutes. (2 marks)

$$\text{Rate of reaction} = \frac{\text{Amount of product made}}{\text{time}} = \frac{15\text{ cm}^3}{5\text{ min}} = 3\text{ cm}^3/\text{min}$$

You can also work out the rate of reaction using the gradient of a graph.

WORKIT!

This graph shows the amount of oxygen being released by catalase against time.

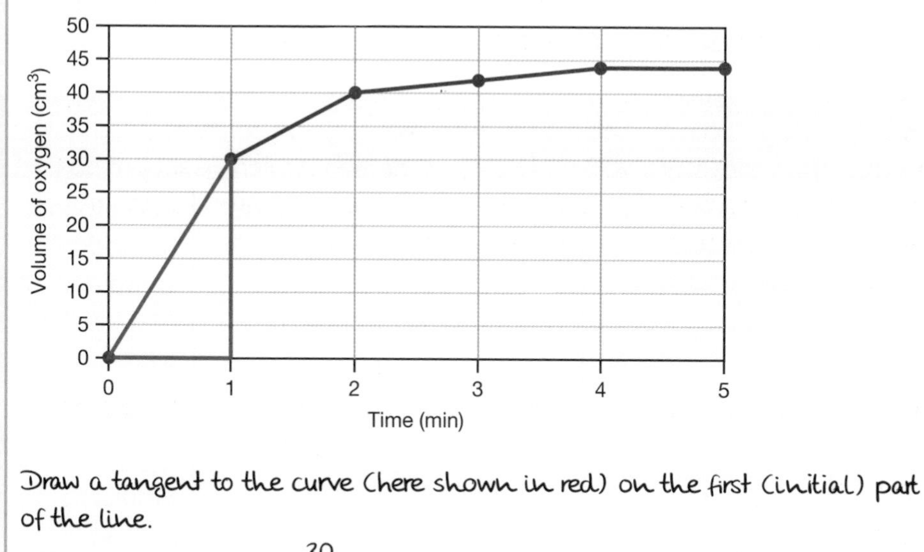

Draw a tangent to the curve (here shown in red) on the first (initial) part of the line.

Initial rate of reaction is $\frac{30}{1}$ = 30 cm^3/min

Draw a line down towards the *x*-axis, and a line across away from the *y*-axis, to form a triangle.

Read off the vertical line and the horizontal line and calculate the gradient.

Here, the vertical line is 30 cm^3, and the horizontal line is 1 min.

The lock and key theory

Enzymes have an active site which is a specific shape. This shape is complementary to the substrate that the enzyme works on. This means that the substrate fits into the active site on the enzyme, like a key fits into a lock. Only the correct substrate can fit into the active site.

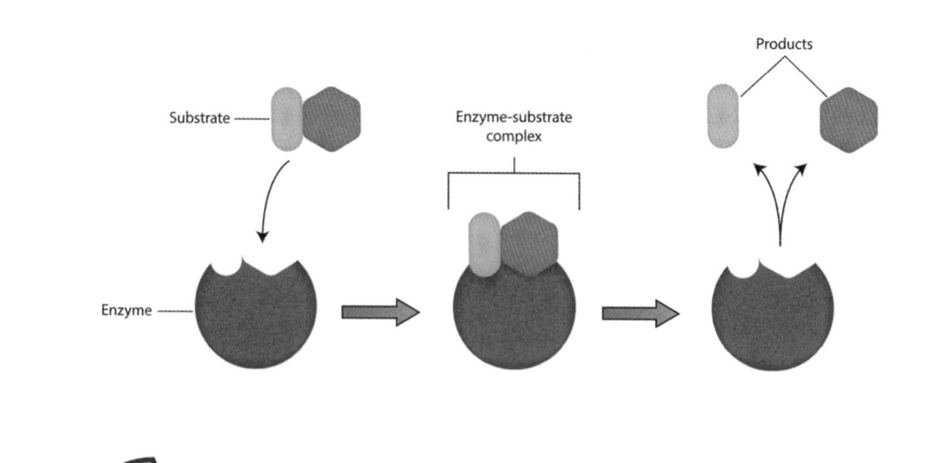

DOIT!

Model the lock and key theory using modelling clay. Use one colour to represent the enzyme, and one colour to represent the substrate.

CHECKIT!

1 What is the effect of placing amylase enzyme into a solution of pH3?

2 20 cm^3 of a product is digested by an enzyme in 4 minutes. Calculate the rate of reaction.

3 Explain why an enzyme is less effective if it is denatured.

Required practical 4: Food tests

For this practical you will need to recall the colours for each of the positive and negative food tests in the table below.

	Benedict's test	Iodine test	Biuret test	Emulsion test
Method	Add Benedict's reagent and heat for several minutes.	Add iodine to food sample.	Add Biuret reagent to food sample.	Add ethanol to food sample, then add distilled water.
Tests for...?	Sugars	Starch	Protein	Lipids
Positive result	Brick red Orange Yellow Green	Blue-black	Lilac	Emulsion formed
Negative result	Blue	Orange	Blue	No emulsion

Practical Skills

In these food tests, you're using qualitative reagents to identify biological molecules.

Describe the results for each test and then explain why these results happened.

WORKIT!

A sample of milk was tested using Benedict's reagent, iodine and Biuret reagent. Describe and explain the results. (3 marks)

The Benedict's reagent turned the milk green, as there is a small amount of sugar in the milk. (1)

The iodine stayed orange, as there is no starch in milk. (1)

The Biuret reagent turned the milk lilac, as there is protein present in milk. (1)

CHECKIT!

1 What could you conclude about a food sample that turned lilac with Biuret reagent?

2 Describe how to carry out the emulsion test.

3 What results would you expect to see if you carried out the iodine test and the Benedict's test on a sample of pasta? Explain your answer.

Required practical 5: The effect of pH on amylase

Amylase is an enzyme found in saliva in the mouth. It digests starch (amylose) into sugars (maltose) and has an optimum pH of 7. In this practical you will need to use your knowledge of enzymes to hypothesize on how pH affects amylase activity. Remember to consider what your control variables will be.

Practical Skills

Investigating the effect of pH

- Use a measuring cylinder or syringe to measure out a volume of amylase and place this into a test tube.
- Use a measuring cylinder or syringe to measure out a volume of buffer of a particular pH and add this to the same test tube.
- Use a measuring cylinder or syringe to measure out a volume of starch solution, add this to the same test tube and set a timer.
- Gradually add iodine to work out when the starch has been digested into sugar. The iodine will be blue-black when starch is present in the test tube, and orange when the starch is no longer present.
- Record results every 30 seconds to make sure your samples are representative.
- Record the time taken for the amylase to digest the starch in buffers of different pH.
- Draw a graph showing the time taken for the amylase to digest the starch against pH.

NAILIT!

The rate of reaction can be worked out by measuring how long it takes for a reactant to be used up, or how long it takes for a product to be made. In this investigation, iodine is used to indicate when the reactant (starch) has been used up.

MATHS SKILLS

Calculating the rate of reaction

To work out the rate of reaction, divide the amount of reactant used, or product made, by time.

WORKIT!

At pH4, 2 cm^3 of starch solution was used up in 140 seconds. Calculate the rate of reaction. (2 marks)

$$\text{Rate of reaction} = \frac{\text{amount of reactant used}}{\text{time}} = \frac{2}{140} = 0.014\ cm^3/s$$

NAILIT!

Remember to always include the units in your answer.

CHECKIT!

1 What does amylase do?

2 Explain what would happen to the rate of reaction of amylase at a pH above pH7.

3 2 cm^3 of starch was used up in 60 seconds. Calculate the rate of reaction.

The heart

> You will not be expected to know the names of the valves!

The human heart is located in the thorax (chest cavity).

The structure and function of the human heart

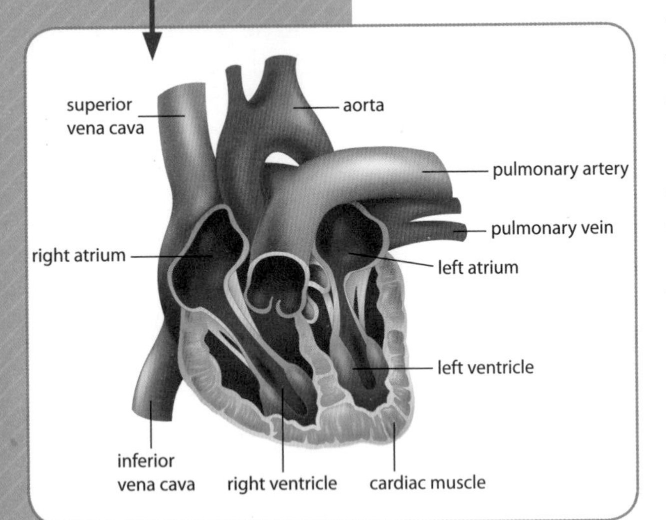

The heart is divided into the right side and the left side. The blood is pumped through both sides at the same time to pump the blood through a double circulatory system.

Deoxygenated blood from the body enters the right atrium of the heart through the superior and inferior vena cava. At the same time, oxygenated blood from the lungs enters the left atrium of the heart through the pulmonary vein. The muscles of the right and left atria contract at the same time, and push the blood into the right and left ventricles. A fraction of a second later, the muscles of the right and left ventricle contract at the same time.

The right ventricle pushes the blood into the pulmonary artery. This blood will now travel to the lungs where gas exchange takes place. The left ventricle pushes the blood into the aorta. This blood will now travel around the body. The blood vessels that supply the heart muscle with oxygen are called the coronary arteries.

Valves between the atria and the ventricles, and inside the pulmonary artery and the aorta, prevent blood from flowing backwards around the heart.

SNAPIT!

Make your own basic diagram of a heart. Take a photo of it and learn the labels.

Pacemakers

A group of cells in the right atrium control the natural resting heart rate. They are the heart's natural pacemaker. If a person has an irregular heart rate, a small electrical device called an artificial pacemaker can be put in place to correct this.

WORKIT!

Describe how oxygenated blood flows through the heart. (4 marks)

Oxygenated blood enters the left atrium (1) through the pulmonary vein. (1)

The muscles in the left atrium contract, pushing the blood into the left ventricle. (1)

When the left ventricle contracts, the blood leaves the heart through the aorta. (1)

CHECKIT!

1 Which vein of the heart carries oxygenated blood? Pulmonary vein

2 What is the purpose of the valves in the heart? Prevent backflows of blood.

3 Explain how the heart maintains a natural resting heart rate and why this is important.

The lungs

The lungs are located in the thorax (chest cavity).

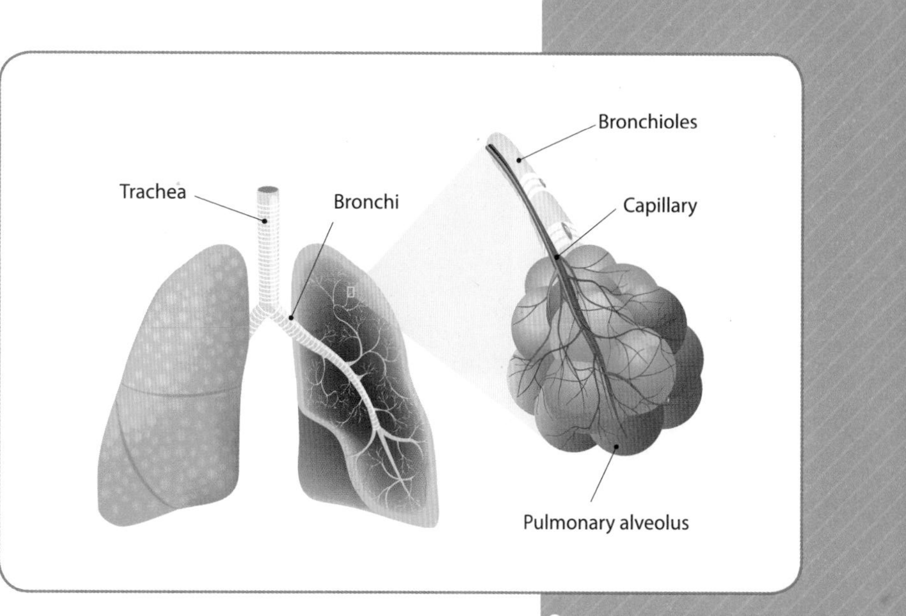

The structure and function of human lungs

The lungs take in air from outside the body for gas exchange. Oxygen is taken in and carbon dioxide is released.

Air travels down the trachea and into the right and left bronchi. The air travels along smaller branches to the alveoli. Each alveolus is supplied with a blood capillary.

Gas exchange takes place in the alveoli. Oxygen diffuses from the alveoli into the blood capillary, and carbon dioxide diffuses from the blood capillary into the alveoli.

Adaptations of the lungs for gaseous exchange

- Many alveoli to increase the surface area.
- The alveoli wall and the capillary wall are one cell thick to provide a short diffusion pathway.
- A good supply of blood capillaries moves oxygen away from the alveoli and maintains a high concentration gradient.

Make your own simple diagram of the lungs and label it.

WORKIT!

Describe how oxygen moves from the trachea and into the blood. (3 marks)

Oxygen moves from the trachea into the right and left bronchi. (1)

Oxygen then moves down smaller branches and into the alveoli. (1)

Oxygen diffuses out of the alveoli and into the blood capillaries. (1)

CHECKIT!

1 Where in the lungs is the concentration of carbon dioxide gas greatest?

2 Why is a short diffusion pathway important in gas exchange?

3 Explain how having a good blood supply helps to increase the rate of diffusion.

Blood vessels

DOIT!

Make cue cards of arteries, veins and capillaries and match the blood vessels to the features and functions.

WORKIT!

Compare and contrast the structure and function of arteries and veins. (4 marks)

Arteries have thick walls, but veins have thin walls. (1)

Arteries have small lumens, but veins have wide lumens. (1)

Arteries carry blood at high pressure, but veins carry blood at low pressure. (1)

Most arteries carry oxygenated blood, but most veins carry deoxygenated blood. (1)

NAILIT!

You could also have mentioned that veins have valves, but arteries do not.

Blood vessels carry blood around the body. There are three types: arteries, veins, and capillaries.

	Artery	Vein	Capillary
	Smooth muscle, Elastic layer, Inner layer	Smooth muscle, Inner layer, Valve	Lumen, Basement membrane, Endothelium
Features	• Thick muscular walls • Small passageways (lumen)	• Thin walls • Wide passageways (lumen) • Valves	• Walls are one cell thick
Function	• Carries oxygenated blood around the body • Carries blood at high pressure	• Carries deoxygenated blood back to the heart • Carries blood at low pressure • Valves prevent the backflow of blood	• Found in the tissues and lungs • Very low blood pressure • Site of gas exchange in tissues and lungs

Calculating blood flow

It is possible to work out the rate of blood flow, if you know the distance the blood travelled and the time that it took.

WORKIT!

Blood flows through a 12 mm artery in 0.2 seconds. What is the rate of blood flow? (3 marks)

$$\text{Rate of blood flow} = \frac{\text{distance travelled by blood}}{\text{time}} = \frac{12}{0.2} = 60 \text{ mm/s}$$

CHECKIT!

1 What type of blood vessel carries deoxygenated blood to the heart?

2 What is the name of the blood vessel that carries oxygenated blood away from the heart?

3 Explain the structure and function of capillaries.

Blood

Blood is a fluid which consists of:

Plasma – a yellow fluid; red blood cells – carry oxygen around the body; white blood cells – protect the body from pathogens; platelets – needed to form scabs if there is a cut; large, insoluble proteins; dissolved substances, such as hormones, oxygen, carbon dioxide.

DO IT!

Practise recognising blood cells by looking at photos of cells in your textbook and online.

Blood cells

There are two types of blood cell: red blood cells and white blood cells.

	Red blood cell	White blood cell
Function	Carry oxygen around the body	Protect the body from pathogens
Adaptation	Biconcave in shape – increased surface area for oxygen to diffuse in No nucleus – more room to pack in haemoglobin Packed with haemoglobin – to carry more oxygen	Phagocytes can change shape – to engulf pathogens Neutrophils have a lobed nucleus – to squeeze through small spaces Lymphocytes have a lot of rough endoplasmic reticulum (with attached ribosomes) – to produce antibodies

WORK IT!

Identify cells A and B in the photograph.

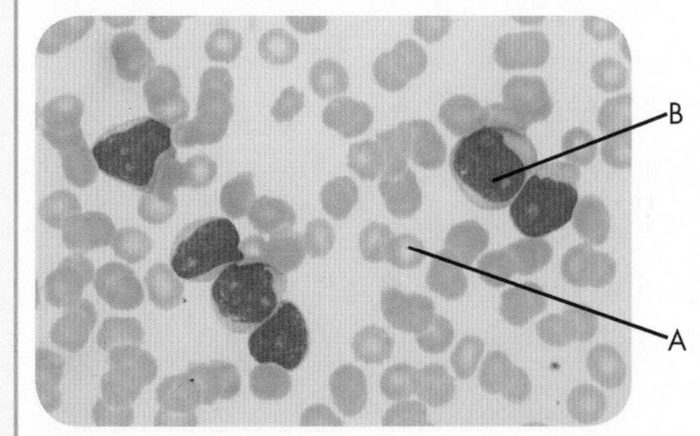

Cell A is a red blood cell. (1)

Cell B is a white blood cell. (1)

NAIL IT!

Red blood cells are smaller than white blood cells, and are red. White blood cells are larger than red blood cells and are usually stained light purple with a pink-purple nucleus.

Practical Skills

Identifying blood cells

You need to be able to recognise blood cells from a drawing and from a photograph taken with a microscope.

CHECK IT!

1 Name two of the main components of blood.

2 How are red blood cells adapted to carry oxygen?

3 Explain why lymphocytes need a large amount of rough endoplasmic reticulum.

Coronary heart disease

Coronary heart disease (CHD) is a non-communicable disease that affects the heart and the coronary arteries.

What causes CHD?

In CHD, layers of fatty material and cholesterol build up inside the coronary arteries (the arteries that supply the heart muscle with oxygen). This makes the passageway (lumen) inside the artery narrower, and reduces the flow of blood.

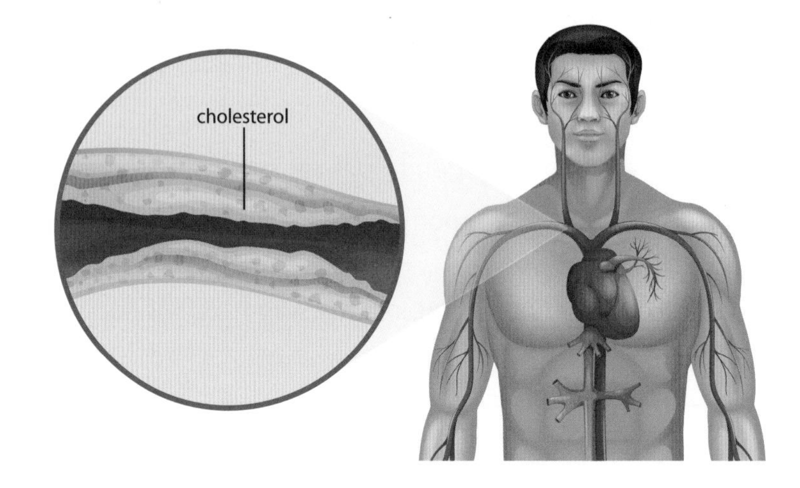

CHD symptoms

There are three main symptoms associated with CHD:

- Angina – chest pains, often brought on by exercise, as the blood supply to the muscles of the heart is restricted.
- Heart failure – muscle weakness in the heart, or a faulty valve, means that the heart does not pump enough blood around the body at the right pressure.
- Heart attack – the blood supply to the muscles of the heart is suddenly blocked, usually by a clot.

Faulty valves

Heart valves are faulty if they:

- Do not open properly – blood cannot pass into the atrium or ventricle.
- Do not close properly – blood can flow backwards around the heart.

In both cases, the heart is put under extra strain and has to pump harder to get the blood through the heart.

STRETCH IT!

Find out about stem cell research into heart and heart valve transplants.

Treatments for CHD

Blocked or restricted coronary arteries can be treated with a stent. This is a small tube that holds the artery open. Medicines called statins may be given. Statins reduce blood cholesterol levels and slow down the rate of fatty material deposit.

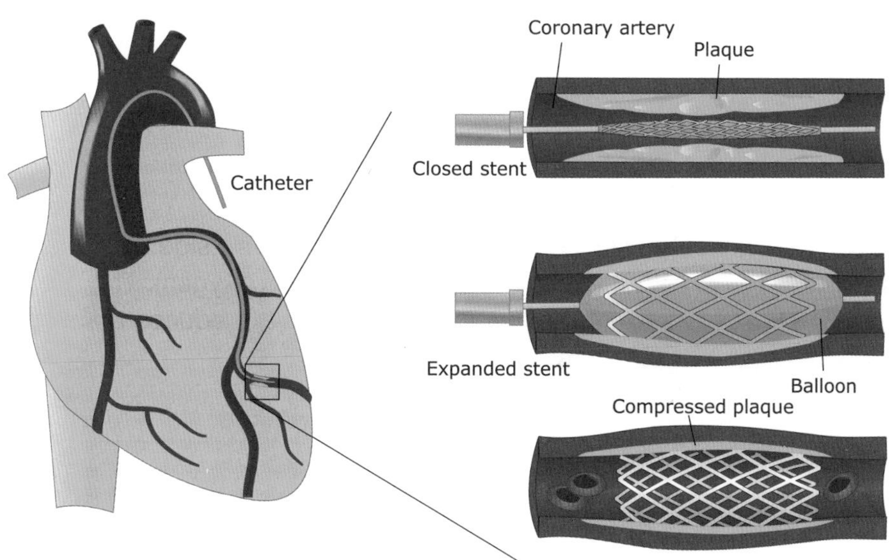

Faulty valves can be treated using a mechanical valve, or a transplanted valve from an animal (usually a pig) or a human donor.

Patients with severe heart failure may have a heart transplant from a human donor, or have an artificial heart while waiting for a transplant. Research is being carried out to make replacement hearts and heart valves from a patient's own stem cells.

NAILIT!

You need to compare the different methods of treatment of CHD and demonstrate the advantages and disadvantages of each method.

DOIT!

Make cue cards for each treatment for CHD and put the advantages and disadvantages of each one on the back. These can be technical, social or ethical reasons.

WORKIT!

Evaluate the methods of treatment for CHD. (4 marks)

Stents hold the coronary arteries open and allow blood flow to the heart, but they are not a permanent solution. Eventually, the artery will need to be bypassed. (1)

Statins reduce blood cholesterol levels but only slow down the rate of fatty material deposit. The fatty material will eventually build up if the patient does not eat a healthy diet. (1)

Faulty valves can be replaced with mechanical or biological valves. Mechanical valves will only last for a while/biological valves may be rejected by the body. (1)

Heart transplants give the patient a new heart but there may be rejection/long waiting list for transplants. (1)

CHECKIT!

1 Name two symptoms of CHD.

2 Describe the causes of CHD.

3 Explain what would happen if a heart valve did not close properly.

Health issues

Health is a state of physical and mental well-being. A person's physical or mental health may be affected by many factors such as diet, stress, genetic background and behavioural choices.

What is a disease?

A disease is a major cause of ill health. There are two types, shown here in the table:

	Communicable	Non-communicable
Cause	Pathogen, such as virus, bacteria, fungi or protozoa	Genetic Lifestyle – diet, stress, exercise levels
Spread	Transmitted from one person or animal to another	Non-transmissible

Diseases that interact

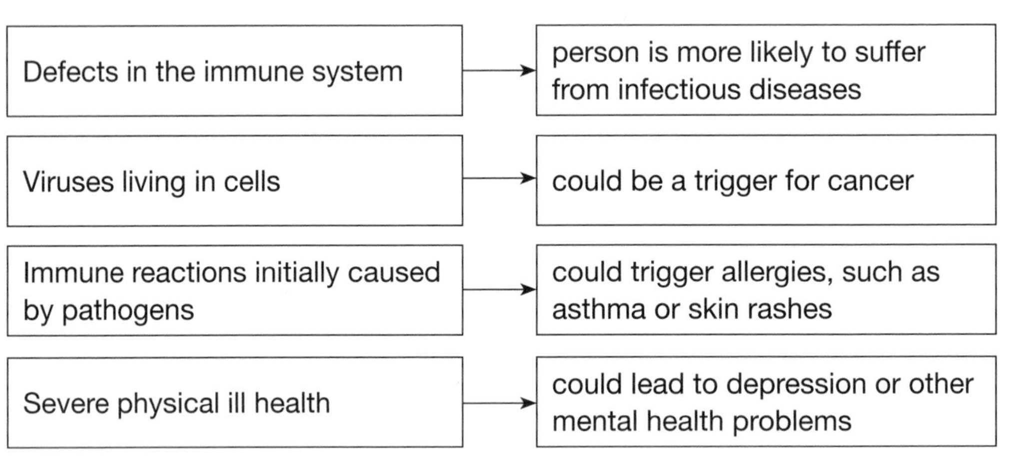

Sampling

Sampling is used to find patterns in peoples' lifestyle choices and their incidence of disease. This is called epidemiological data. For example, epidemiological data has shown that people with diets high in saturated fat are more likely to suffer from coronary heart disease.

MATHS SKILLS

Scatter diagrams and correlations

You should be able to identify a correlation between two variables by looking at a scatter diagram.

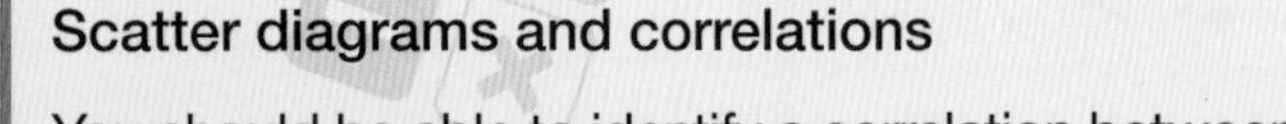

These scatter diagrams show the three possible types of correlation:

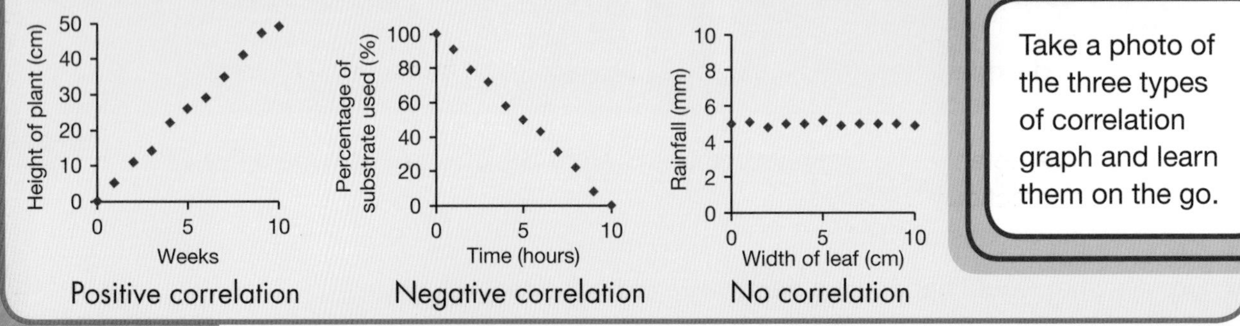

Positive correlation Negative correlation No correlation

SNAP IT!

Take a photo of the three types of correlation graph and learn them on the go.

WORKIT!

This graph shows the percentage of males and females diagnosed with diabetes at different ages. Describe the pattern using data from the graph. (3 marks)

In all age groups, except for age 16-34, more males than females were diagnosed with diabetes. (1)

The percentage of females diagnosed with diabetes increased from 1% at age 16-34, to 13% at age 75+. (1)

The percentage of males diagnosed with diabetes increased from 0.5% at age 16-34, to 17.5% at age 75+. (1)

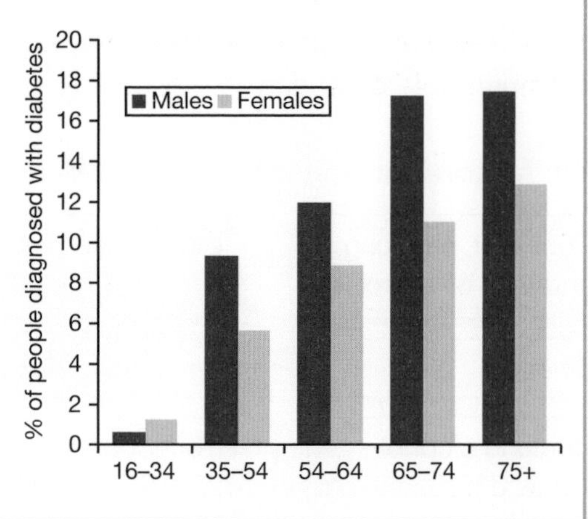

WORKIT!

In a class of 30 students, 6 ate five portions of fruit and vegetables per day, 8 ate four portions, 10 ate two portions, 3 ate one portion and 2 ate no portions. One student ate six portions.
Draw a frequency table to show this data. (2 marks)

Portions of fruit and vegetables	Frequency
0	ll
1	lll
2	𝍸 𝍸
3	0
4	𝍸 lll
5	𝍸 l
6	l

NAILIT!

Bar charts show data from distinct categories and the bars do not touch. Histograms show numerical data and the bars touch.

MATHS SKILLS

Translating disease incidence information

Make sure you can write numerical data using information from a graph. You should also be able to draw a graph from numerical data.

DOIT!

Look at health data graphs online and practise interpreting them.

CHECKIT!

1 In an experiment, more product was produced the longer the reaction was carried out. A graph was plotted of mass produced against time. State the type of correlation shown.

2 Describe and explain how you would show the data in the frequency table above as a graph.

3 Suggest how defects in a person's immune system could make them more likely to suffer from infectious diseases.

Effect of lifestyle on health

NAILIT!

The human and financial costs of non-communicable diseases:

Human cost
- Premature death
- Pain and discomfort
- Limited mobility
- Unable to work

Financial cost
- Cost of medicines
- Cost of hospital treatments
- Lost working days – loss in income/ productivity

Lifestyle and substances in the environment can lead to an increase in non-communicable diseases.

Risk factors for non-communicable diseases

Poor diet, smoking and lack of exercise	→ increased risk of cardiovascular disease
Obesity	→ increased risk for type 2 diabetes
Drinking alcohol	→ effect on liver and brain function
Smoking	→ increased risk of lung disease and lung cancer
Smoking and alcohol	→ effect on unborn babies
Carcinogens, such as ionising radiation	→ increased risk of cancer

Many diseases are caused by the interaction of a number of factors.

WORKIT!

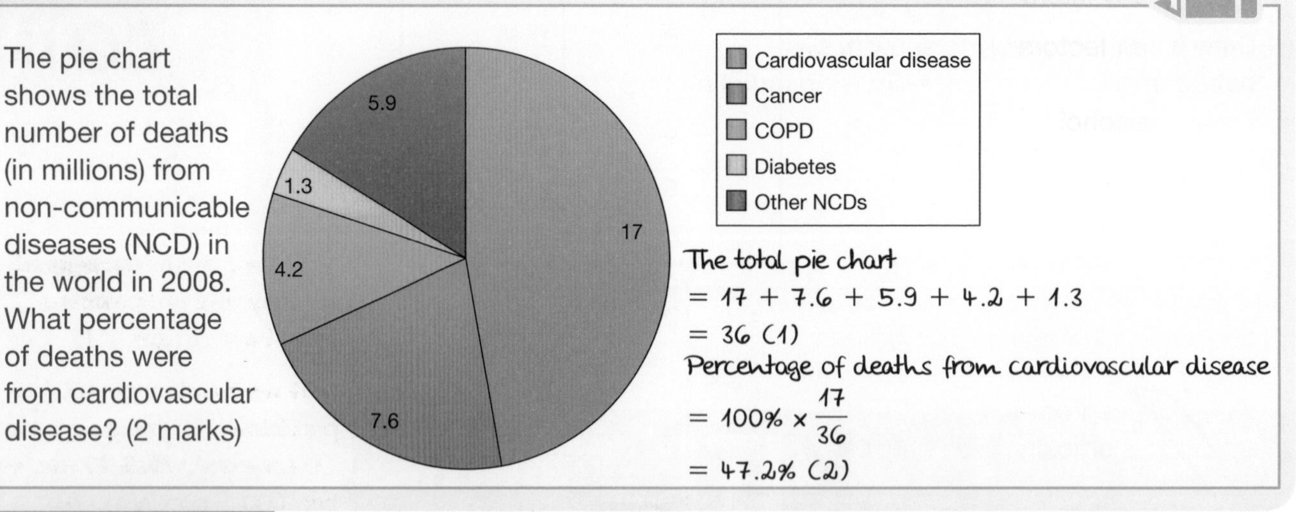

The pie chart shows the total number of deaths (in millions) from non-communicable diseases (NCD) in the world in 2008. What percentage of deaths were from cardiovascular disease? (2 marks)

The total pie chart

$= 17 + 7.6 + 5.9 + 4.2 + 1.3$

$= 36$ (1)

Percentage of deaths from cardiovascular disease

$= 100\% \times \frac{17}{36}$

$= 47.2\%$ (2)

CHECKIT!

1. Name two risk factors associated with cardiovascular disease.
2. State one financial cost of living with a non-communicable disease.
3. The number of deaths worldwide from cardiovascular disease in 2030 is predicted to be 23.6 million. Using data from the pie chart above, calculate the percentage increase in deaths this would be since 2008.

Cancer

Cancer is a non-communicable disease with many interacting risk factors.

What is cancer?

Mutations in cells are caused by a number of different factors. When normal cells divide, they are checked for mutations. If any mutations exist, the cell is not able to go through mitosis (a type of cell division, see page 18). Also when normal cells divide, they lie next to each other in a uniform way. Cancer cells are abnormal. They divide without being checked, and form a mass, lying over the top of one another. A mass of cancer cells is called a tumour.

Types of tumour

There are two types of tumour:

Benign – growths of abnormal cells, contained in one area, usually within a membrane. They do not invade other parts of the body.

Malignant – growths of abnormal cells that invade neighbouring tissues and spread through the blood to different parts of the body, where they form secondary tumours.

Causes of cancer

There are many interacting risk factors associated with cancer, including:

- genetic risk factors
- smoking
- drinking alcohol.
- poor diet
- ionising radiation

STRETCH IT!

Research how checkpoints in the cell cycle (see page 18) check the cell's DNA for mistakes and prevent uncontrolled cell division.

NAIL IT!

Describe the pattern using data from the graph.

WORK IT!

This graph shows the annual deaths from lung cancer in Australia per 1 000 in 2015, and the number of cigarettes smoked per day. Describe the evidence from the graph that shows that smoking may cause lung cancer.

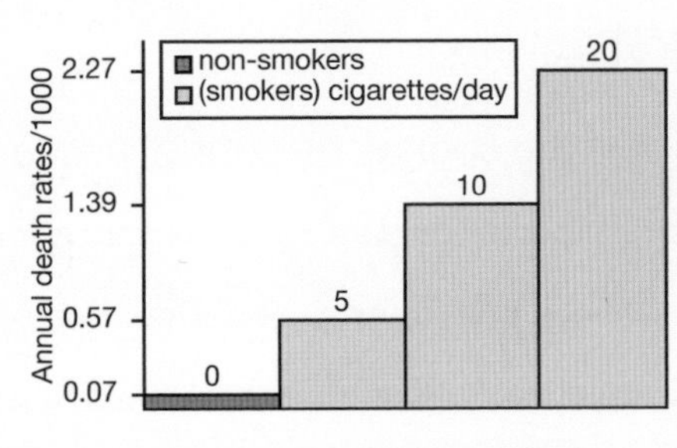

The more cigarettes smoked per day, the more deaths from lung cancer. (1)

0.07 non-smokers per 1000 of population died from lung cancer (1) compared with 2.27 smokers per 1000 of population who smoked 20 cigarettes per day. (1)

CHECK IT!

1. Name two risk factors for developing cancer.
2. Describe how a tumour is formed.
3. Suggest why malignant tumours are more difficult to treat than benign ones.

Plant tissues

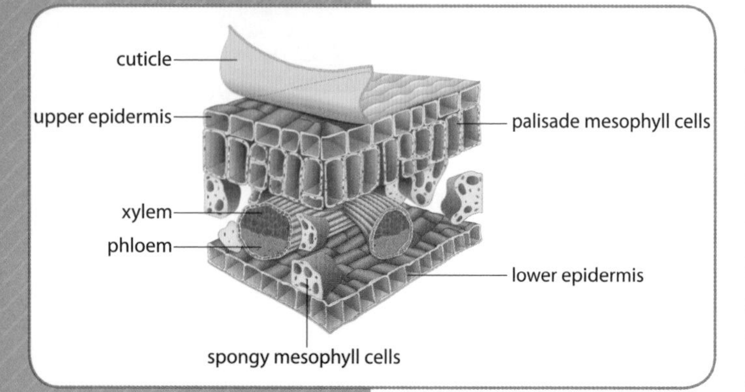

Different parts of plants are made from different types of tissue.

The leaf

The leaf is a plant organ. It is made of several plant tissues: epidermal tissues, palisade mesophyll, spongy mesophyll, xylem and phloem. The lower epidermal tissues contain guard cells. These control the opening and closing of small holes called stomata. This is where carbon dioxide and oxygen enter and exit the leaf.

NAILIT!

The diagram above shows a transverse section of a leaf. You may be asked to observe and draw this.

DOIT!

Look at a transverse section of a leaf using a microscope and identify the different tissues. Practise drawing the leaf.

Function of plant tissues

Plant tissue	Function
Epidermal tissues	Allow light to reach the palisade tissues – thin and transparent
Palisade tissues	Light absorption for photosynthesis – tightly packed, column shaped cells, packed with chlorophyll
Spongy mesophyll	Site of gas exchange – gases dissolve in thin layer of water on the surface of the loosely packed cells, and then diffuse into air spaces
Xylem	Transport of water to the leaf – narrow, hollow tubes with a waterproof layer in the cell walls
Phloem	Transport of sugar sap around the plant – narrow, almost hollow tubes
Meristem tissue	Divides rapidly at the root and shoot tips to provide new plant cells – can become any type of plant cell

WORKIT!

Explain how the spongy mesophyll tissue is adapted for its function.

The cells are loosely packed to allow space for the gases to diffuse. (1)

The cells have a thin layer of water on the surface for the gases to dissolve into. (1)

The cells are close to the lower epidermis/guard cells/stomata so that the gases can diffuse in and out of the leaf. (1)

NAILIT!

In Q2, limit your drawing to 4-5 cells.

CHECKIT!

1 Which plant tissue transports water to the leaf?

2 Draw and label the palisade tissue.

3 Explain why it is important that the upper epidermis is thin and transparent.

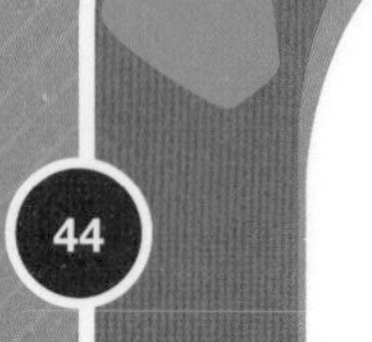

Transpiration and translocation

Transpiration is the loss of water from the top part of the plant. Translocation is the movement of sugar sap up and down the plant.

What is transpiration?

The root, stem and leaves form a plant organ system for transport of substances around the plant. In transpiration, water moves into the root hair cells and into the xylem by osmosis (see page 23). Water travels up the xylem using the forces of adhesion and cohesion, until it reaches the leaf. In the leaf, water diffuses by osmosis into the cells, and then evaporates onto the surface of the spongy mesophyll cells. Water vapour then diffuses out of the leaf through the stomata.

What is translocation?

Photosynthesis in the leaves produces sugars. Some of these are needed in respiration, and to make molecules for the plant. The rest is stored in the roots as an energy store.

Sugar sap is made of dissolved sugars, and moves up and down the plant in the phloem tissue. Phloem tissue is made of many phloem cells joined end to end. The end plate of each phloem cell has many small pores in it and is called a sieve plate. The phloem cells join together to make a long narrow, almost hollow tube. Mature phloem cells are alive, but only have a limited number of sub-cellular structures.

Tissues in transpiration

1. Root hair cells are adapted to take in water by having a long hair to provide a large surface area (see page 10). The cytoplasm inside the root hair cell is a concentrated solution compared to the dilute solution in the soil, so water moves into the cells by osmosis. Mineral ions are also taken up by the root hair cells, by active transport (see page 26). Root hair cells have many mitochondria to provide the energy from respiration to do this.
2. Xylem tissue is made of many xylem cells joined end to end. The end plate of each xylem cell is removed, to make a long narrow, hollow tube for water and dissolved mineral ions to travel up. The movement of water through the xylem is called the transpiration stream.

   Mature xylem cells are dead. They do not contain a nucleus or any sub-cellular structures. The cell walls of the xylem contain lignin, which makes the cells waterproof. This helps the water to stay inside the xylem. Lignin also gives strength to the xylem and therefore the stem of the plant.
3. Guard cells form pairs on the lower epidermis of the leaf. When there is plenty of water, the cells become turgid and allow the stomata to open. When there is little water, the walls of the guard cells push together to close the stomata. This controls the loss of water from the leaf. Gases can also enter and exit the cell through the open stomata.

DO IT!

Make a mind map on transpiration and another on translocation, using all of the main points and then adding more detail.

NAIL IT!

You will not be expected to describe the phloem tissue or translocation in detail.

DO IT!

Paint a leaf with clear nail varnish and then look at the guard cells and stomata using a microscope.

Measuring the rate of transpiration

The rate of transpiration can be measured by measuring the amount of water taken up by the plant. The simplest way to do this is to weigh the plant, and then weigh the plant again after a few hours. The change in mass is the mass of the water that has left the plant.

Increasing the rate of transpiration

The rate of transpiration can be increased by:

- increasing temperature
- increasing light levels
- increasing air movement
- decreasing humidity.

WORKIT!

The rate of transpiration of a plant was found by measuring the loss of water over four hours. The investigation was carried out at 20°C, 40°C, and 60°C. Which temperature do you think had the fastest rate of transpiration? Justify your answer. (4 marks)

60°C (1) because the rate of transpiration is higher with increased temperatures. (1)

The water molecules will have more kinetic energy at higher temperature so rate of diffusion is higher (1) and more water could diffuse out of the leaves. (1)

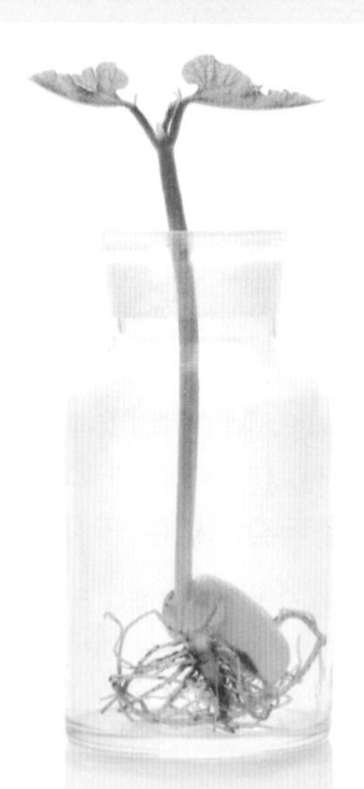

CHECKIT!

1. Describe the process of transpiration.
2. Name two factors that increase the rate of transpiration.
3. In an investigation, 12 cm^3 of water was lost by transpiration in 4 hours. Calculate the rate of transpiration.

Tissues, organs and organ systems

REVIEW IT!

1 a Name two enzymes found in the human digestive system.
  b What are carbohydrates digested into?
  c Describe how an enzyme works.
  d What would be the effect of placing an enzyme with an optimum pH of 7 into a solution with a pH of 3?

2 Describe the food tests for the following food groups:
  a Sugars b Starch c Protein.

3 a i Which blood vessel carries deoxygenated blood to the heart?
    ii Which blood vessel carries oxygenated blood to the heart?
  b Describe how the blood is kept flowing through the heart in the right direction.
  c How is the heartbeat controlled?

4 a Describe the pathway of the air from the mouth to the site of gaseous exchange in mammals.
  b Describe how the gaseous exchange surface in mammals is adapted.
  c Blood flows through a 10 mm section of capillary in the lungs in 0.4 seconds. Calculate the rate of blood flow.

5 A patient with CHD has previously had a heart attack.
  a What causes a heart attack?
  b The doctor advises the patient to either take statins or have a stent fitted in their coronary artery. Evaluate the advantages and disadvantages of each treatment.

6 a In plant tissues, guard cells open and close to form small pores called stomata. What is the function of the stomata?
  b An investigation into the number of open stomata at different temperatures was carried out. The results are shown in the table below:

Temperature	Number of open stomata			Mean number of open stomata
	1	2	3	
10	14	16	12	14
20	34		38	36
30	49	51	44	
40		78	81	82

    i Complete the table.
    ii Correct any mistakes in the table.
    iii Describe what the data shows.
    iv Suggest a reason for these results.
  c Increased humidity decreases the rate of transpiration. What factors will increase the rate of transpiration?

Infection and response

Communicable diseases

SNAPIT!

Take a photo of this table and learn one or two examples of each type of pathogen.

Communicable diseases are caused by pathogens that are spread from one person or animal to another.

Pathogens

Pathogens are microorganisms that cause disease. Not all microorganisms are pathogens.

There are four groups of pathogens:

Pathogen	Appearance	Transmission	Diseases
Bacteria	Single-celled Prokaryotic	Airborne – coughing and sneezing Direct contact with infected person/animal/ sharp object In food and water	Tuberculosis Tetanus Cholera *Salmonella* food poisoning
Viruses	Much smaller than a bacterium Needs a host	Airborne – coughing and sneezing Direct contact with infected person	Common cold Influenza HIV
Fungi	Single-celled Eukaryotic Chitin cell wall	Airborne – carried on wind Direct contact with infected person/animal Indirect contact – towels and changing room floors	Ringworm Athlete's foot
Protists	Single-celled Eukaryotic No cell walls	Vector – carried by insect	Malaria Dysentery

Bacteria and viruses may reproduce rapidly inside the body. Bacteria may produce poisons (toxins) that damage tissues and make us feel ill. Viruses live and reproduce inside cells, causing cell damage.

Reducing the spread of diseases

The spread of diseases can be reduced by:

- regular handwashing
- using tissues
- vaccination programmes
- preparing food safely
- access to clean drinking water
- clean, well-ventilated homes
- practising safe sex
- reducing the population of some insects in infected areas, e.g. mosquitoes.

WORKIT!

In some areas of Brazil, large quantities of insecticide (chemicals that kill insects) are being used to kill populations of *Anopheles* mosquitoes. Explain how this will reduce the cases of malaria. (4 marks)

Malaria is caused by a protist/protozoa/plasmodium (1) carried by the *Anopheles* mosquito. (1)

The protist/protozoa/plasmodium gets into the human body when the mosquito bites them. (1)

Fewer mosquitoes mean that people are less likely to get bitten. (1)

STRETCHIT!

Find out about research into releasing genetically modified mosquitoes into infected areas.

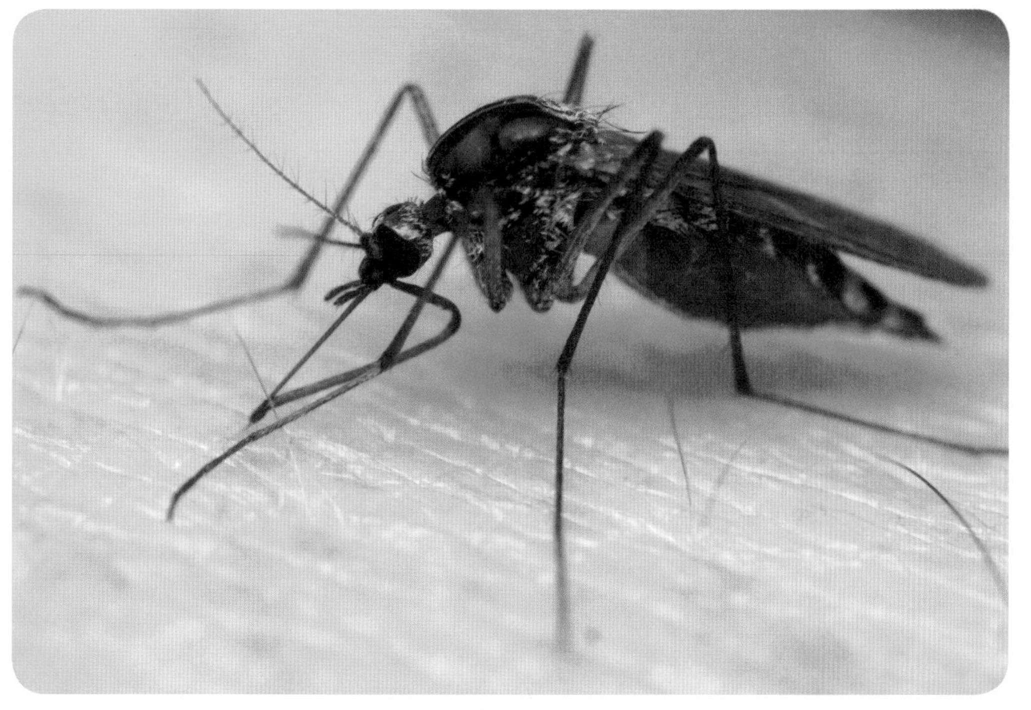

An *Anopheles* mosquito

CHECKIT!

1. Name two diseases caused by a virus.
2. Describe how bacteria may be transmitted.
3. Suggest why people living in damp, crowded housing may be more susceptible to communicable diseases.

NAILIT!

Think about pathogens and insects that thrive in damp conditions, and how most viruses and bacteria can be spread.

Viral diseases

STRETCH IT!

Find out why measles has been on the rise in the UK for the past few years.

NAIL IT!

HIV is caused by contact with infected body fluids. It can be passed from mother to child.

DO IT!

Put these viral infections into a table with the headings: virus, causes, symptoms, treatment. This will be easier to remember than a page of text.

NAIL IT!

In your answer to Q3, make sure you include any benefits to the other plants, but also what the infected plant may leave behind in the soil.

Viruses live and reproduce inside cells, causing cell damage. Some viruses can cause serious illnesses and even death if not treated quickly.

Measles

Measles is spread by breathing in droplets from sneezes and coughs. The symptoms of measles are a red skin rash and a fever. Measles is a serious illness and can result in death if complications arise. Most babies and young children in the UK are vaccinated against measles. This means that they are unlikely to catch the measles virus.

HIV

Human immunodeficiency virus (HIV) is transmitted through sexual contact, or contact with infected blood, such as when drug users share needles. The virus causes a flu-like illness in the first instance. Untreated, HIV attacks the cells of the immune system, developing into acquired immune deficiency syndrome (AIDS). This is where the immune system no longer functions, and the person may die from other infections or cancer. HIV can be treated with antiretroviral drugs, which control the progression of the disease.

Tobacco mosaic virus (TMV)

TMV is a widespread plant virus that affects many types of plant, including tobacco and tomato plants. The virus appears as a black 'mosaic' pattern on the leaves. This affects plant growth, as the dark areas cannot carry out photosynthesis. TMV spreads from plant to plant through direct contact, or in infected soil. TMV is treated by removing infected plants, and washing hands before planting new plants.

WORK IT!

Describe how the spread of HIV could be reduced. (3 marks)

Practise safe sex. (1)

Free needle exchanges so there is no need for drug users to share needles. (1)

Check blood products for HIV to prevent transmission through blood transfusions. (1)

Give pregnant women with HIV medicines to prevent transmission of HIV to the fetus. (1)

CHECK IT!

1 Describe how measles is spread.

2 Describe the symptoms of untreated HIV.

3 Describe the effect of removing plants infected with TMV from a field.

Bacterial diseases

Salmonella food poisoning and gonorrhoea are both caused by bacteria.

Salmonella food poisoning

Salmonella food poisoning is caused by the bacterium *Salmonella enterica*, in food prepared in an unhygienic way. The symptoms of this type of food poisoning are fever, abdominal cramps, vomiting and diarrhoea. The salmonella bacteria move into the digestive system and secrete toxins that cause the symptoms. Most people recover without treatment, but in severe cases, people may need to go into hospital to be treated for dehydration.

The way to prevent salmonella food poisoning is to prepare and store food in a hygienic way, and make sure that food is cooked thoroughly. Salmonella can be found in poultry (chickens and turkeys) and eggs, so in the UK, poultry are vaccinated against salmonella.

Gonorrhoea

Gonorrhoea is a sexually transmitted disease (STD) caused by the bacterium *Neisseria gonorrhoeae*. The symptoms are a thick yellow or green discharge from the vagina or penis, and pain when urinating. Untreated, gonorrhoea can lead to severe complications and in rare cases, infertility or septicaemia.

Gonorrhoea can be treated with antibiotics, but new strains of *Neisseria gonorrhoeae* have become resistant to some antibiotics, such as penicillin. In the UK, there are currently two antibiotics that can be used to treat this infection. The spread of gonorrhoea can be controlled by practising safe sex, for example, using a condom.

NAILIT!

You will not be expected to recall the full names of the bacteria.

STRETCHIT!

Find out about the highly drug-resistant 'super-gonorrhoea' that is spreading across the UK.

DOIT!

Condense your notes on these bacteria to a few words for each to remember for your exams.

WORKIT!

A person wants to avoid getting Salmonella food poisoning. What advice would you give to them? (4 marks)

Do not eat undercooked eggs or poultry, or drink unpasteurised milk. (1)

Wash your hands with soap and water before handling food. (1)

Keep food properly refrigerated. (1)

Clean cooking surfaces before preparing food. (1)

CHECKIT!

1 Which type of medicine is used to treat bacterial infections?

2 Explain why poultry in the UK are vaccinated against Salmonella.

3 Describe and explain the consequences of antibiotic resistance in treating gonorrhoea.

Fungal and protist diseases

NAILIT!

You will not be expected to recall the full names of the fungi or protist.

STRETCHIT!

Find out about the life cycle of the malaria protist.

NAILIT!

Mosquitoes are part of a food chain. The food chain will be disrupted if the mosquitoes are removed.

Fungi and protists can cause a range of diseases, including rose black spot and malaria.

Rose black spot

Rose black spot is a fungal disease, spread by water or wind, that affects roses. It is caused by the fungus, *Diplocarpon rosea*, and infects the leaves of the rose with black or purple spots. This affects photosynthesis, as the areas of the leaf covered in spots cannot photosynthesise. This reduction in photosynthesis will affect the growth of the plant. Infected leaves turn yellow and drop to the ground.

Rose black spot can be treated with fungicides (chemicals that kill fungi), or removing and destroying the infected leaves.

Malaria

Malaria is caused by a type of protist. These protists live in the saliva of the female *Anopheles* mosquito and are spread to humans when the mosquito bites them. The mosquito is a vector. The protists spend some of their life cycle in the human body, and move back into the mosquito when the mosquito bites an infected person. In this way, the protists can be spread from person to person.

Malaria causes a severe fever, which can occur several times over a person's lifetime, and can be fatal. Malaria can be treated with antimalarial drugs. The main way to prevent getting malaria is to prevent yourself being bitten by a mosquito, using mosquito nets at night and long-sleeved clothes or insect repellent during the day. Many people also take antimalarial drugs as a preventative. Mosquitoes like to breed in small pools of water. Another way of preventing the spread of malaria is to prevent the mosquitoes from breeding by removing pools where they breed, or killing mosquitoes using pesticides (chemicals that kill pests).

WORKIT!

Compare the advantages and disadvantages of removing mosquitoes from an area. (4 marks)

Removing the mosquitoes will reduce the chances of getting malaria. (1)

Removing the mosquitoes will remove a food source in the food chain. (1)

Some animals may decrease in number due to the lack of mosquitoes (1) but it will save some human lives. (1)

CHECKIT!

1 Which type of medicine is used to treat malaria?

2 How do mosquito nets help to prevent malaria?

3 Describe how to prevent the spread of rose black spot.

Human defence systems

NAILIT!

Think about non-specific and specific defences.

The human body has many different types of defence against pathogens.

Non-specific defence systems

These non-specific defences try to stop all pathogens getting into our bodies.

	Defence
Skin	Physical barrier against pathogens Breaks in the skin form scabs Sweat glands produce sweat that inhibits pathogens
Nose	Small hairs and mucus trap airborne particles
Trachea and bronchi	Mucus traps pathogens and is moved up to the throat by small hairs called cilia
Stomach	Hydrochloric acid and protease enzymes in the stomach kill pathogens

DOIT!

To get a better idea of how phagocytosis happens, look at a video online of phagocytes engulfing pathogens.

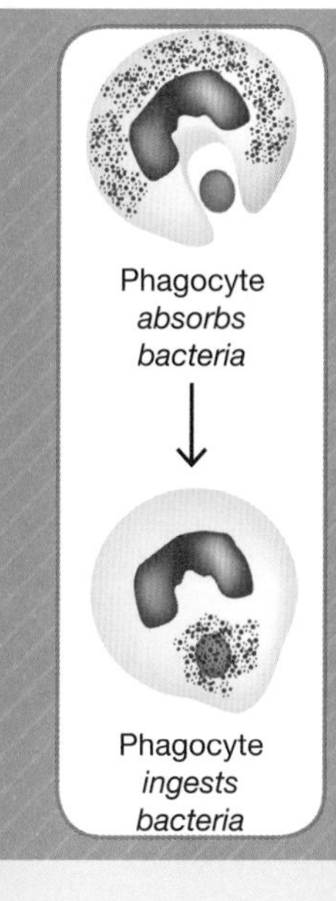

Specific immune systems

If pathogens get past the non-specific defences, there are two types of white blood cell that can destroy them.

1 *Phagocytes*

These are a type of white blood cell that can engulf (take inside their body) any type of pathogen. They do this by extending their cell membrane around the pathogen until it is surrounded. The pathogen then enters the phagocyte and is digested by enzymes. This process is called phagocytosis.

2 *Lymphocytes*

These are a type of white blood cell that attack specific pathogens. They produce antibodies that have several functions:

- attach to pathogens and prevent them from entering cells
- attach to pathogens and target them for phagocytosis
- act as antitoxins by attaching to toxins.

WORKIT!

Describe and explain the body's defences against the influenza virus. (4 marks)

The influenza virus cannot pass through the skin, which is a non-specific barrier. (1)
Mucus traps any inhaled virus, which is then moved up to the throat by cilia. (1)
Phagocytes will engulf the virus by phagocytosis. (1)
Lymphocytes will produce antibodies against the virus. (1)

CHECKIT!

1 Name two non-specific defences against pathogens.

2 Outline the process of phagocytosis.

3 Explain how the specific immune system stops bacterial toxins making a person feel ill.

Vaccination

Vaccinations are given to people to prevent them from developing illnesses from pathogens.

How does vaccination work?

Vaccination is the introduction into the body of small quantities of dead or inactive forms of a pathogen into the body. These dead or inactive pathogens stimulate the white blood cells to make antibodies. If the same pathogen re-enters the body, white blood cells respond quickly to make the correct antibodies, and prevent illness.

Vaccination programmes

In the UK, most babies and young children are vaccinated against a number of pathogens. The more people who are vaccinated against a pathogen, the less likely the pathogen is to spread through the population. This is called herd immunity. There will always be a few people who cannot be vaccinated. They will be protected by herd immunity and are unlikely to become ill from that pathogen.

Around the world, vaccination programmes are used to prevent the spread of disease. Vaccination programmes successfully rid the world of smallpox in 1979.

DO IT!

Write out a step-by-step of how vaccinations work.

NAIL IT!

You do not need to know details of vaccination schedules or side effects associated with specific vaccines.

STRETCH IT!

Find out about global vaccination programmes to reduce the spread of tuberculosis (TB).

WORK IT!

What is herd immunity? (2 marks)

When most of a population is vaccinated against a pathogen. (1)
Non-vaccinated people are protected from infection with that pathogen. (1)

CHECK IT!

1 What is in a vaccine?

2 Suggest why the UK no longer routinely vaccinates against TB.

3 In 1998, a doctor falsely claimed that the MMR vaccine caused autism. Using data from the graph describe and explain the percentage of children in the UK vaccinated with the MMR vaccine and the incidence of measles.

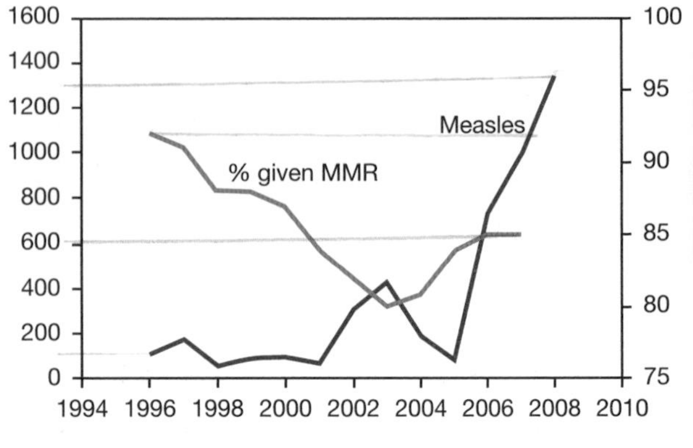

Antibiotics and painkillers

Antibiotics

Antibiotics, such as penicillin, are medicines that destroy or inhibit the growth of bacteria. They are used to treat bacterial infections, such as kidney infections, or diseases, such as pneumonia. Antibiotics can be given as pills, a cream, or an injection. Specific bacteria are killed by specific types of antibiotic, so it important to be prescribed the correct one. They cannot be used to treat any diseases caused by a virus, such as influenza.

Antibiotic resistance

Since antibiotics were first used in the 20th century, the number of deaths from infectious bacterial diseases has greatly reduced. However, many bacterial strains are resistant to some antibiotics. Some bacterial strains, such as MRSA, are resistant to most antibiotics. Antibiotic resistance is increasing and this is concerning as it may lead to increased deaths. Methods to reduce antibiotic resistance are:

- to be careful with how antibiotics are prescribed
- to take the full course of prescribed antibiotics
- to discover new antibiotics.

Painkillers and other medicines

Painkillers do not kill pathogens, but can be used to treat the symptoms of disease, such as headache and fever.

Antiviral drugs are medicines that are used to treat diseases caused by a virus. However, this is difficult because when viruses invade the body, they move inside the cells. Any medicine that kills the virus would also have to attack the body's own cells, and this could cause damage inside our bodies.

DO IT!

Make a table for antibiotics, antivirals, and painkillers, outlining what each one does.

STRETCH IT!

Find out about MRSA and antibiotic resistance.

NAIL IT!

If asked a question like the one below, make sure you use percentages from the graph in your answer to Q3.

WORK IT!

The graph shows the percentage of *Acinibacter*, a type of bacteria that are resistant to the antibiotic, imipenem, in different years. What data from the graph shows that antibiotic resistance is increasing? (3 marks)

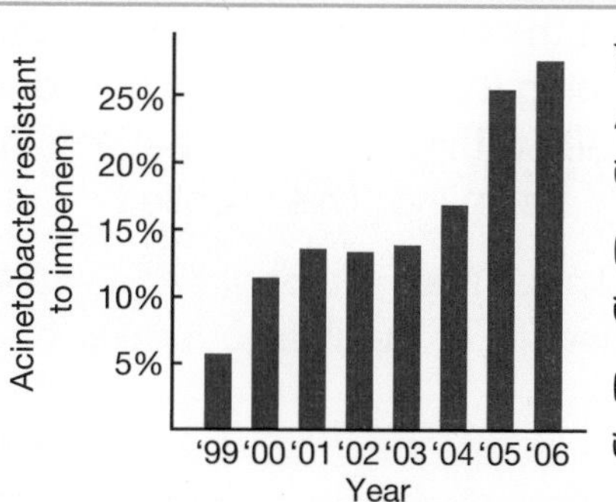

From 1999 to 2005, the percentage of Acinibacter that are resistant to imipenem increases. (1)

In 1999, 6% of Acinibacter are resistant to imipenem. (1)

In 2005, 27% of Acinibacter are resistant to imipenem. (1)

CHECK IT!

1. Give an example of an antibiotic.
2. Explain why antibiotics should not be given to treat a viral disease.
3. Calculate the percentage increase in the antibiotic resistance of *Acinibacter* to imipenem from 1999 to 2005.

New drugs

DO IT!

Read about Alexander Fleming and how he discovered penicillin from some dirty plates!

Sometimes clinical trials don't go to plan. Find out about an incidence when the clinical trial had to be stopped early.

WORKIT!

Explain why new drugs need to be tested before being given to patients. (3 marks)

To make sure that the drug is not toxic/poisonous. (1)

To check the efficacy/how well the drug works. (1)

To find the correct dosage for the drug. (1)

New drugs are discovered every year. They may be extracted from plants or microorganisms, or synthesised in a laboratory.

Discovering new drugs

Traditionally, new drugs were discovered by extracting compounds from plants or microorganisms. For example:

- The painkiller, aspirin comes from willow bark.
- The heart drug, digitalis comes from foxgloves
- Penicillin comes from the *Penicillium* mould.

New drugs are still discovered in this way, but chemists in the pharmaceutical industry can also alter already existing compounds to make more effective medicines.

Testing new drugs

Before a new drug can be given to a patient, it must be tested for safety and effectiveness.

Step 1: preclinical testing – test new drug on cells, tissues and animals in a laboratory. The drug is tested for toxicity (is it poisonous?), efficacy (how well does it work?) and dose (how much of it do we need?). If the drug is toxic or not effective it does not pass to step 2.

Step 2: clinical trials – a very low dosage of the new drug is given to healthy volunteers and patients. If the drug is safe, it moves to step 3.

Step 3: clinical trials – different doses of the new drug are given to healthy volunteers and patients to find the optimum dose for the drug.

Step 4: wider clinical trials – the new drug is given to patients and the efficacy monitored. Often the drug is compared to another commonly used drug, or a placebo (something that looks like the new drug but contains no medicine). To test the efficacy without bias, the patients and the doctor are not told which drug they are taking. This is called a double blind trial.

The results of clinical trials are peer reviewed (judged by other scientists) and published in scientific or medical journals.

CHECKIT!

1 How are new drugs discovered?

2 Describe what happens during preclinical testing.

3 Why is it important to avoid bias during clinical trials?

Monoclonal antibodies

Monoclonal antibodies are made by white blood cells called lymphocytes.

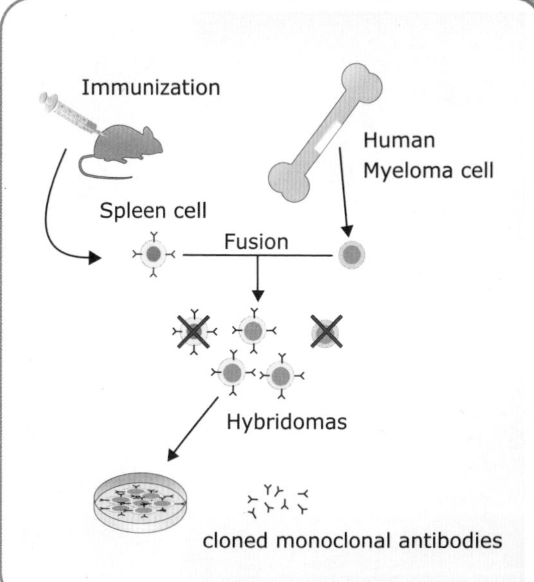

How are monoclonal antibodies made?

Monoclonal antibodies are made by stimulating mouse lymphocytes.

1. The mouse is injected with an antigen. An antigen is a foreign substance that triggers an immune response in the body.
2. This stimulates one of the mouse lymphocytes to form many clones of itself.
3. These cloned lymphocytes make antibodies that are complementary to the antigen. All of the monoclonal antibodies are identical. → are found in the B-cells
4. The lymphocytes are combined with a type of tumour cell to make a hybridoma cell. → because they can't divid
5. Hybridoma cells can divide and make the antibody.

This method can be used to make a large amount of monoclonal antibodies. These antibodies can then be collected and purified.

Make a copy of the diagram above, and take a photo to revise from later.

What do monoclonal antibodies bind to?

Monoclonal antibodies have a specific 'Y' shape that binds only to the binding site on the antigen. The antigen is usually a protein. The shape of the antibody is complementary to the shape of the binding site. This means that the antibodies cannot bind to any other antigen, and can be used to target a particular protein, chemical, or cell in the body.

Explain how monoclonal antibodies can be made to target a particular cell in the body. (4 marks)

An antigen from the cell is injected into a mouse. (1)

Mouse lymphocytes are stimulated to make many monoclonal antibodies that are complementary to the cell's antigen. (1)

The lymphocytes are combined with a tumour cell to make a hybridoma. (1)

The hybridoma makes many monoclonal antibodies that can target the cell. (1)

CHECKIT!

H1 What type of cell produces antibodies?

H2 Describe how the monoclonal antibody binds to the antigen.

H3 Suggest what would happen if monoclonal antibodies from a mouse were injected into a rabbit.

Monoclonal antibody uses

H

Monoclonal antibodies have many scientific and medical uses.

How are monoclonal antibodies used?

Monoclonal antibodies are used for:

- Diagnosis – in pregnancy tests.
- Blood testing – to measure the levels of hormones/chemicals in blood.
- Disease detection – to test for the presence of pathogens.
- Research – locate or identify specific molecules in a cell by binding them with a fluorescent dye.
- To treat diseases – monoclonal antibodies are bound to a radioactive substance, drug or chemical and used to treat cancer. The antibodies bind specifically to the cancer cells and do not harm the other cells of the body.

Side effects of monoclonal antibodies

The most common side effect of using monoclonal antibodies is an allergic reaction. Other side effects are chills, fever, rashes, feeling sick, and breathlessness.

DO IT!

Put each of these uses of monoclonal antibodies onto a cue card and test your friends on how many they can remember!

NAIL IT!

You will not be expected to describe any specific tests.

WORK IT!

Evaluate the advantages and disadvantages of using monoclonal antibodies. (4 marks)

Monoclonal antibodies are very specific/they will only bind to a particular antigen, so they can be used to detect molecules. (1)

Will only bind to cancer cells, so can deliver medicines without harming healthy cells. (1)

May cause discomfort to the animals that are being used to make monoclonal antibodies. (1)

Using monoclonal antibodies in treatments may cause side effects. (1)

NAIL IT!

Remember to include ethical issues and potential harm in your answer to the WorkIt!

CHECK IT!

H1 Name two uses of monoclonal antibodies.

H2 What is the advantage of using monoclonal antibodies to treat cancer cells?

H3 Outline how monoclonal antibodies can be used to detect a protein in the blood.

Plant diseases

H

Plant diseases are caused by pathogens or ion deficiencies.

Detecting plant diseases

Diseases can be detected in plants by looking at their growth and the colour of their leaves.

Condition	Common diseases
Stunted growth	Viral infection
Spots on leaves	Rose black spot
Areas of decay/rotting	Blight
Growths	Cankers/galls
Malformed stems or leaves	Blight
Discolouration	Tobacco mosaic virus
Presence of pests	Aphid infection

You do not need to know all of these diseases; only black spot, TMV and aphid infection.

Identification of plant diseases

Plant diseases can be identified by referring to a gardening manual or website. Infected plants can also be taken to a laboratory for identification of the pathogen. Pathogens can be detected using kits containing monoclonal antibodies (see page 58).

Common plant diseases

Three common plant diseases are black spot, tobacco mosaic virus, and aphid infection.

Black spot

This disease affects roses, and is caused by a fungus, *Diplocarpon rosae* (see page 52). The disease appears as purple or black spots on the upper surface of the rose leaves.

Tobacco mosaic virus (TMV)

This viral disease was first identified in tobacco plants, but affects many different species of plant, including tomatoes (see page 50). It can be identified as a black 'mosaic' pattern on the leaves.

Aphid infection

Aphids are small green insects that feed on the sap of plants. They can infect many different types of plant. An aphid infection stunts the growth of the plant and the leaves may be curled or distorted.

Copy out this table and take a picture of it. You need to learn by heart the effect of nitrate and magnesium deficiencies.

Ion deficiencies in plants

Plant growth may be stunted, or the leaves may be discoloured, if the plant is deficient in mineral ions.

Mineral ion	Needed for	Deficiency
Nitrate	Making proteins	Poor growth, yellow leaves
Magnesium	Making chlorophyll for photosynthesis	Yellow leaves, chlorosis

Chlorosis

Chlorosis is the yellowing of the leaves, caused by disease, lack of sunlight, or mineral ion deficiency. In chlorosis, the chlorophyll pigment, which makes the leaves green, is no longer being produced. If it is a mineral ion deficiency, adding magnesium to the soil will return the leaves to normal.

WORKIT! H

A plant has black spots on its leaves. How can you identify what is causing the spots? (3 marks)

Look in a gardening manual/website. (1)

Take the plant to a laboratory for testing. (1)

Use a test kit containing monoclonal antibodies. (1)

CHECKIT!

1 What type of pathogen causes Rose black spot?

2 Explain which mineral a plant is deficient in, if it has yellow leaves and poor growth.

3 This graph shows the number of tobacco plants in a field, infected by TMV over 5 years. Describe and explain the trend, and suggest what the farmer can do to reduce the number of plants infected.

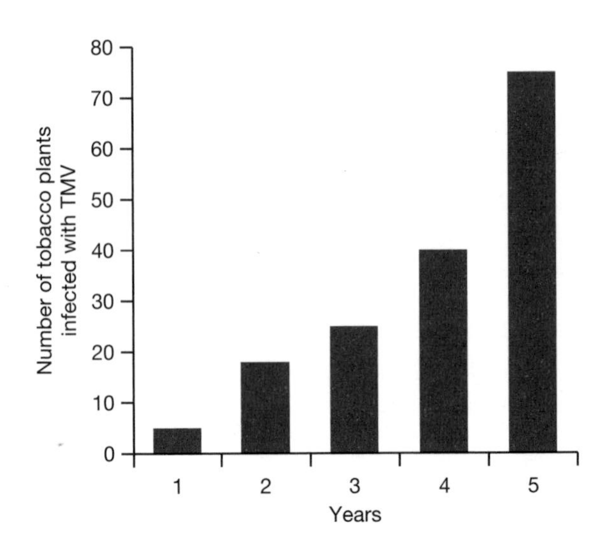

Plant defences

Plants have a number of physical and chemical responses to pathogens.

Physical plant defences

- All plant cells have cellulose cell walls. These are strong, and prevent pathogens from entering the plant through the stems.
- Trees have a thick layer of dead cells called bark on the surface of the trunk. These contain a waterproof substance called lignin which acts as barrier to pathogens through the trunk.
- Leaves have a waxy cuticle on the top surface which is thick and tough. This prevents pathogens from entering the plant through the leaf.

Chemical plant defences

Plants, such as mint, produce antibacterial chemicals to defend themselves against pathogens. Plants also protect themselves against herbivores (animals that eat plants) by producing poisons. Stinging nettles, for example, have small needle-like cells on their leaves that break off and inject irritating toxins when touched.

Mechanical adaptations

The structure of the plant can help to protect it from herbivores:

1 Thorns – on rose bushes and brambles, protect from animals.

2 Hairs – on leaves protect the plant from insects.

3 Drooping or curling leaves – mimosa leaves rapidly droop down when touched to protect them from herbivores.

4 Mimicry – making the plant's leaves or flowers look like something else. For example, the *Lithops weberi* has leaves that look exactly like stones.

STRETCH IT!

Find out how we use some antibacterial chemicals from plants in our everyday lives.

WORK IT!

Explain how the structural defences of a plant protect it from pathogens. (3 marks)

Pathogens cannot pass through the lignin in the bark on the trunk of a tree. (1)

Pathogens cannot pass through the cellulose cell wall of plant cells. (1)

The waxy cuticle on the upper surface of a leaf prevents pathogens from entering the leaves. (1)

DO IT!

Have a look at some photos of *Lithops weberi* online.

CHECK IT!

1 Give two examples of a chemical defence in plants.

2 Explain how mimicry protects a plant from herbivores.

REVIEW IT!

Infection and response

1 a What is a pathogen?
  b Name two communicable diseases that are caused by bacteria.
  c Describe how bacterial diseases are transmitted.
  d What type of drugs can be used to treat bacterial infections?

2 a Compare the structure of bacteria and viruses.
  b Measles is a virus that is often caught by small children.
    i Describe the symptoms of measles.
    ii Identify some ways in which measles could be prevented.
  c TMV is a virus that affects plants.
    i What does TMV stand for?
    ii Explain how TMV affects the growth of plants.

3 The graph below shows the number of deaths from malaria from 2000 to 2015.

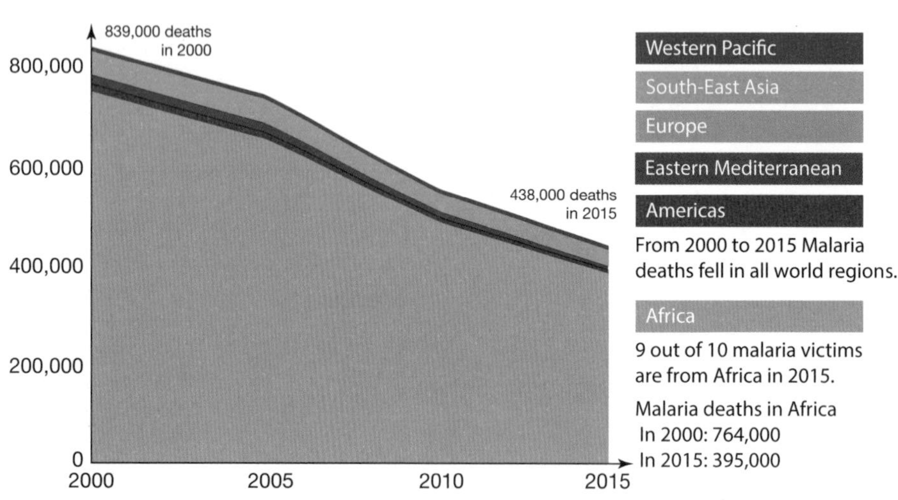

  a What type of microorganism causes malaria?
  b i Describe the pattern of the graph.
    ii Suggest reasons for this pattern.

4 a Describe two non-specific defences the body has against pathogens.
  b Compare two types of white blood cell that defend the body against pathogens.

5 A new antibiotic (C) is being tested in a laboratory, along with two other antibiotics (A and B) that are in common use. The results are shown in the table below:

Antibiotic	Number of bacteria killed			Mean number of bacteria killed
	1	2	3	
A	34	35	37	35
B	44	39	42	42
C	140	126	132	133

  a i Describe the results of the table.
    ii Suggest a reason for these results.
  b Explain why it is important to keep making new antibiotics.

Photosynthesis

Photosynthesis is the process of using energy from sunlight to make sugars.

The process of photosynthesis

Photosynthesis takes place in the leaves. Chloroplasts in the cells of the leaf are packed with a pigment called chlorophyll. Chlorophyll absorbs sunlight and uses it to convert carbon dioxide and water into glucose (a type of sugar) and oxygen. The word equation for this is:

$$\text{carbon dioxide} + \text{water} \xrightarrow{\text{light}} \text{glucose} + \text{oxygen}$$

The chemical equation for photosynthesis is:

$$6CO_2 + 6H_2O \xrightarrow{\text{light}} C_6H_{12}O_6 + 6O_2$$

This is an endothermic reaction because more energy is taken in by the reaction than is given off. The light energy from the Sun is converted into chemical energy in the plant.

Reactants for photosynthesis

The carbon dioxide needed for photosynthesis enters the leaf through the stomata. The water gets to the leaf through the xylem (see page 45).

Products of photosynthesis

Product of photosynthesis	
Glucose	Used in respiration Makes other important molecules such as starch, fats and amino acids
Oxygen	Used in respiration Released from the plant through the stomata

DO IT!

Write out the word equation and the chemical equation until you have learned them by heart.

WORK IT!

Explain the effect on photosynthesis if a plant is left in a bright room but not watered. (3 marks)

Photosynthesis cannot occur. (1)

The plant would only have carbon dioxide and sunlight. (1)

The plant needs water to carry out photosynthesis. (1)

NAIL IT!

Look at the chemical formula to see how many of each atom you have.

CHECK IT!

1. What are the products of photosynthesis?
2. Explain where the concentration of chlorophyll would be highest in the leaf.
3. The chemical formula for glucose is $C_6H_{12}O_6$. Explain where the carbon atoms have come from.

Rate of photosynthesis

The rate of photosynthesis changes depending on a number of different, interacting factors.

Factors that affect the rate of photosynthesis

The rate of photosynthesis increases if:

- the temperature increases (up to an optimum temperature)
- light intensity increases
- carbon dioxide concentration increases
- the amount of chlorophyll increases.

Limiting factors

The rate of photosynthesis can be measured by measuring the volume of oxygen produced by a plant in a certain time, under certain conditions. The shape of the graph for each condition is shown below:

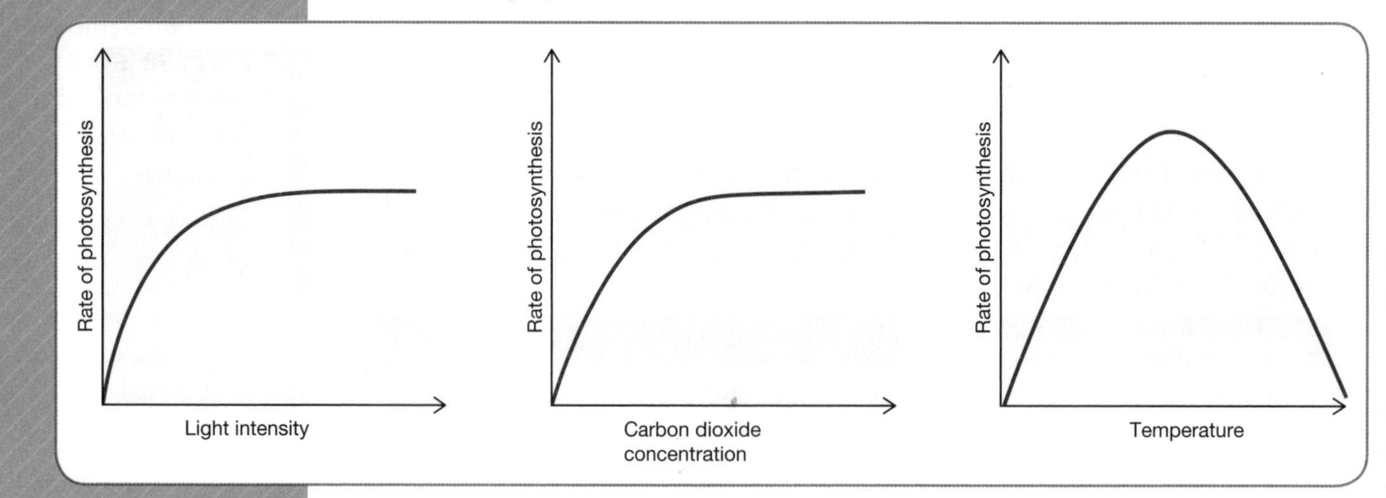

SNAP IT!

Copy out these graphs and take a photo to help you learn their shapes.

Effect of light and carbon dioxide

As the light intensity, or the carbon dioxide concentration increases, the rate of photosynthesis increases, until a certain point. After this point, the rate of photosynthesis remains constant. That is because one of the other factors is limiting the rate of photosynthesis. This is called the limiting factor. For example, in the first graph, as the light intensity increases and rate of photosynthesis increases, light intensity is the limiting factor. When the rate of photosynthesis no longer increases some other factor (such as the concentration of carbon dioxide) has become the limiting factor.

Effect of temperature

As the temperature increases, the rate of photosynthesis increases, until the optimum temperature for the enzymes and proteins that carry out photosynthesis is reached. Above this temperature, the enzymes and proteins begin to denature and can no longer carry out their function. Therefore the rate of photosynthesis decreases.

Measuring the rate of photosynthesis

In an experiment, the volume of oxygen given off by a plant is measured over time. The rate of photosynthesis can be calculated by dividing the volume of oxygen collected over time:

$$\text{Rate of photosynthesis} = \frac{\text{volume of oxygen}}{\text{time}}$$

WORKIT!

$12\,cm^3$ of oxygen is collected in 3 minutes. Calculate the rate of photosynthesis. (3 marks)

Rate of photosynthesis $= \frac{12\,m^3}{3\,min} = 4\,cm^3/min$ (3)

NAILIT!

Remember to include units when you answer a question.

H

The inverse square law and light intensity in photosynthesis

In an investigation into the effect of light on the rate of photosynthesis, a lamp is moved away from the plant. As the distance from the plant increases, the light intensity decreases. The light intensity is inversely proportional to the square of the distance.

H

Increasing the rate of photosynthesis

Farmers need to increase the rate of photosynthesis in plants in order to increase the yield (the volume of plants, fruit or vegetables produced). They do this by:

- growing plants in a greenhouse to increase the temperature
- increasing the hours of light by switching lights on at night
- increasing the amount of carbon dioxide inside the greenhouse.

STRETCHIT!

Read up on the greenhouses at Thanet Earth to get an idea of how photosynthesis is controlled to increase the yield of tomatoes.

CHECKIT!

1 Explain the term 'limiting factor'.

2 Describe and explain what would happen to the rate of photosynthesis if the carbon dioxide concentration were decreased.

3 Calculate the rate of photosynthesis if $28\,cm^3$ of oxygen is collected in 4 minutes.

There are many methods for carrying out this investigation and a diagram in the exam may not be the same as the one you used.

Find two alternative methods for investigating the rate of photosynthesis.

Required practical 6: Investigating the effect of light intensity on the rate of photosynthesis

In this practical you will need to use your knowledge of photosynthesis to hypothesize about how light intensity will affect the rate of photosynthesis.

The rate of photosynthesis at different intensities of light can be calculated by measuring the volume of oxygen given off by the plant in a certain time.

Practical Skills

Select suitable apparatus to collect the oxygen gas given off by a piece of pondweed. You will need to:

- make sure that the stem of the pondweed is cut underwater, to prevent any air bubbles getting into the stem
- make sure that the oxygen gas cannot escape from the glassware
- make sure there is only one source of light
- keep all variables, except for light intensity, constant.

MATHS SKILLS

Calculating the rate of photosynthesis

The rate of photosynthesis can be calculated by dividing the volume of oxygen released by the time taken.

WORKIT!

The data from the investigation is shown in the table below.

Calculate the rate of photosynthesis at each distance. (5 marks)

Distance of lamp from plant (cm)	Volume of oxygen released (cm^3)	Time (min)	Rate of photosynthesis
0	200	5	200/5 = 40 cm^3/min
10	200	5	200/5 = 40 cm^3/min
20	100	5	100/5 = 20 cm^3/min
30	60	5	60/5 = 12 cm^3/min

NAILIT!

Think about any stage of your investigation that was not as accurate as it could have been. How could you improve the reliability and validity of your investigation?

CHECKIT!

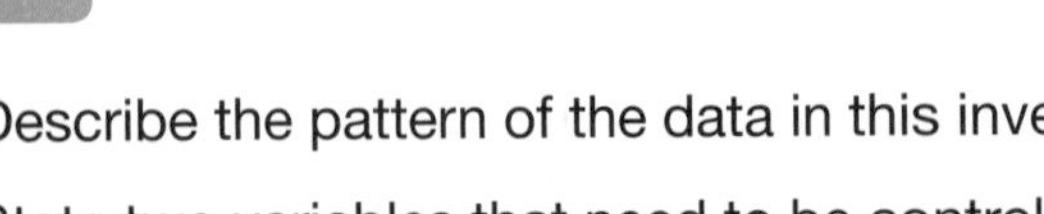
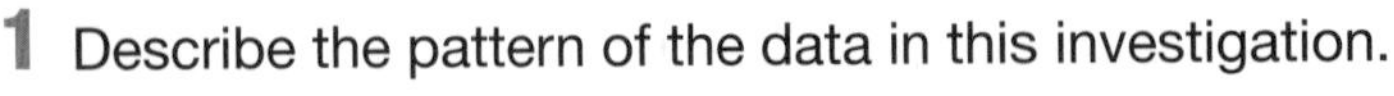

1 Describe the pattern of the data in this investigation.

2 State two variables that need to be controlled in this investigation.

3 Suggest how you could improve this investigation.

Uses of glucose

Glucose is a type of sugar. It is made by the process of photosynthesis, and has the chemical formula, $C_6H_{12}O_6$.

How is glucose used?

More glucose is made by photosynthesis than is needed, so it can be stored as starch or oil to be used later. Starch and oils can be broken down into glucose by enzyme action.

Use of glucose	Animals	Plants
Used in respiration to make energy	✓	✓
Converted into insoluble glycogen for storage	✓	
Converted into cellulose, which strengthens the cell wall		✓
Converted into insoluble starch for storage		✓
Used to produce fat for storage	✓	✓
Used to produce oil for storage		✓
Used to produce amino acids for protein synthesis		✓

To make proteins, plants also need to take up nitrates from the soil.

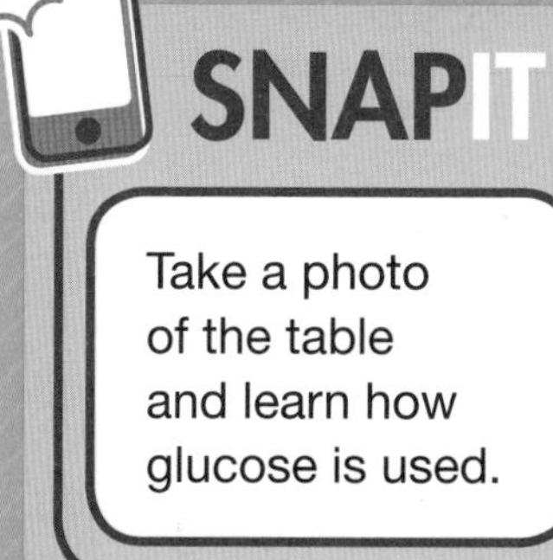

Take a photo of the table and learn how glucose is used.

WORKIT!

Compare the way in which glucose is used in animals and plants. (4 marks)

Glucose is used for respiration in both animals and plants. (1)

Glucose is used for making amino acids in plants. (1)

Glucose is stored as glycogen in animals, but as starch in plants. (1)

Glucose is used to produce fat for storage in animals, but usually used to produce oils for storage in plants. (1)

NAILIT!

Think back to digestion (page 28). Enzymes are used to digest starch into sugars, such as glucose.

CHECKIT!

1 Where in a plant is glucose made?

2 What is the source of glucose for animals?

3 Explain the importance of storing glucose as glycogen or starch.

Respiration

Respiration happens in all living cells. It is the process of converting glucose into energy.

The process of respiration

Respiration is an exothermic reaction, because it releases more energy than is taken in. The chemical energy in glucose is converted into energy for the cell to:

- carry out chemical reactions to build larger molecules
- move
- keep warm.

STRETCH IT!

Find out how respiration occurs inside the mitochondria.

Aerobic respiration

Aerobic respiration uses oxygen. It takes place in the cytoplasm of the cell, and in the mitochondria. The word equation for aerobic respiration is:

glucose + oxygen → carbon dioxide + water

The chemical equation for aerobic respiration is:

$$C_6H_{12}O_6 + 6O_2 \rightarrow 6CO_2 + 6H_2O$$

Aerobic respiration is carried out most of the time in animals and plants and releases a lot of energy, compared to anaerobic respiration.

Anaerobic respiration

Anaerobic respiration takes place without oxygen. It only takes place in the cytoplasm of the cell and does not involve the mitochondria. Muscle cells carry out this type of respiration when they are exercising quickly, for example, when a person is sprinting. It does not release very much energy, as the glucose is not completely oxidised, and can only be carried out in humans for a short time.

The word equation for anaerobic respiration in muscles is:

glucose → lactic acid

The lactic acid builds up in the muscle cells, causing muscle fatigue and cramp. The lactic acid is broken down using oxygen after the exercise is finished. This is called the oxygen debt.

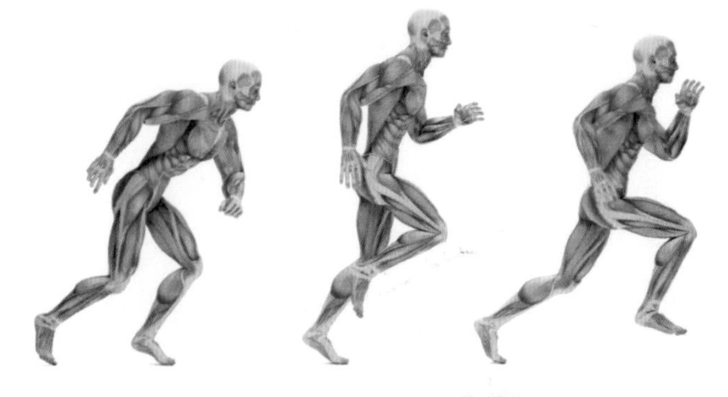

Plant and yeast cells can carry out anaerobic respiration for a much longer time, and produce different products. This process is called fermentation.

The word equation for anaerobic respiration in yeast and plants is:

glucose → ethanol + carbon dioxide

Uses of fermentation

Product of fermentation	Use
Carbon dioxide	Causes bread and cakes to rise
Ethanol	Making beer/wine/spirits

DO IT!

Learn these three word equations and one chemical equation by heart.

WORK IT!

Compare the processes of aerobic and anaerobic respiration. (4 marks)

Aerobic respiration uses oxygen and anaerobic respiration does not use oxygen. (1)

Aerobic respiration releases more energy than anaerobic respiration. (1)

In humans, aerobic respiration produces carbon dioxide and water, whereas anaerobic respiration produces lactic acid. (1)

In yeast, aerobic respiration produces carbon dioxide and water, whereas anaerobic respiration produces carbon dioxide and ethanol. (1)

1 Where in the cell does aerobic respiration take place?

2 Describe how yeast is used to make wine and bread.

3 The graph below shows yeast being grown anaerobically at different temperatures. Describe and explain the pattern.

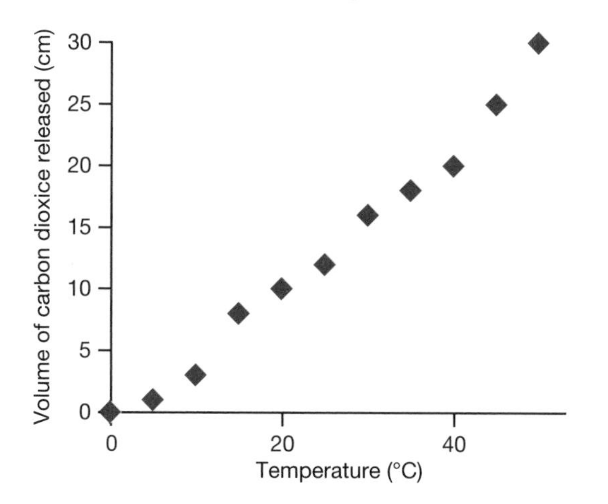

Response to exercise

The human body responds to the increased demand for energy during exercise.

The effect of exercise

During exercise, the muscle cells require more energy in order to keep moving. This means that the rate of cellular respiration must increase. In order to do this, the cells need more oxygen and glucose. These are supplied to the cell by:

- an increase in heart rate so that the blood flows to the cells more quickly
- an increase in the breathing rate to oxygenate the blood more quickly
- an increase in breath volume to take in more oxygen with each breath.

DO IT!

Test your pulse before and after one minute of vigorous exercise. Test your pulse again after five minutes rest. Has it returned to normal?

Anaerobic exercise

When the body exercises vigorously, the cells may respire anaerobically. This means that no oxygen is used, and there is a build up of lactic acid in the muscles (see page 68) due to the incomplete oxidation of glucose. After a short time, the muscles will be fatigued and stop contracting efficiently.

Practical Skills

Investigations into exercise

The heart rate can be measured by measuring the pulse rate. Count the number of pulses in 15 seconds and multiply the number by four. This gives the beats per minute.

H

The oxygen debt

Lactic acid is removed from the cells and transported through the blood to the liver. The liver uses oxygen to convert the lactic acid into useful products, such as glucose. This process requires extra oxygen and is called the oxygen debt.

H

WORK IT!

A person goes for a five minute jog. They measure their pulse rate before and after exercising, and find their pulse rate has increased. Explain why. (3 marks)

The muscle cells need to respire more quickly (1)
to provide more energy for the muscle cells to move. (1)
The heart beats faster to get oxygen and glucose to the muscle cells more quickly. (1)

CHECK IT!

1. A person counts 12 pulses in 15 seconds. What is their heart rate?
2. Explain why the breathing rate increases during exercise.
3. Explain why sprinters need to breathe deeply after finishing a race.

Metabolism

Metabolism is all of the enzyme-controlled reactions in the body. The energy for these reactions comes from respiration in the cells.

Synthesis and breakdown of molecules

There are two types of metabolic reaction.

1 Complex molecules are made out of simpler ones. For example:

sugars → complex carbohydrates
amino acids → proteins
3 fatty acids + glycerol → lipids

These reactions are important in the body for growth, repair and energy storage.

2 Complex molecules are broken down into simpler ones. For example:

complex carbohydrates → sugars
proteins → amino acids
lipids → 3 fatty acids + glycerol

These reactions are important in digestion, to make the molecules small enough to be absorbed into the bloodstream in the small intestine.

DO IT!

Make revision cards of each metabolic reaction and learn some examples.

Metabolism reactions

Here are some other important metabolic reactions:

- respiration
- in plants – formation of amino acids from glucose and nitrate ions
- in animals – breakdown of excess proteins to urea for excretion in urine.

WORK IT!

Explain why metabolism needs energy. (3 marks)

Energy is needed to break the bonds in a large complex molecule. (1)

Enzymes need energy to carry out a reaction. (1)

Energy is needed to form new bonds between smaller molecules. (1)

CHECK IT!

1 Give an example of a metabolic reaction.

2 Explain why the production of complex molecules is important in the body.

3 Suggest why respiration is described as a metabolic reaction.

NAIL IT!

Look back at the process of respiration on page 68.

REVIEW IT! Bioenergetics

1 a What is the chemical formula of glucose?
  b Name two uses of glucose in the human body.
  c Compare how glucose is stored in plants and animals.
2 a Explain why during exercise the breathing rate, breath volume, and heart rate increase.
  b Describe how you could measure the pulse rate.
  c Explain what is meant by an oxygen debt after anaerobic exercise.
3 Respiration is a metabolic reaction.
  a Give one further example of a metabolic reaction.
  b i What is the word equation for aerobic respiration?
    ii What is the balanced symbol equation for aerobic respiration?
4 a What are the products of photosynthesis?
  b Name two factors that increase the rate of photosynthesis.
  c What is meant by the term 'limiting factor'?
5 Some students investigated the rate of photosynthesis at different light intensities.
  a i Draw the expected shape of the graph.
    ii Label the axes correctly.
  b Explain the shape of the graph you have drawn.
6 An investigation into the rate of photosynthesis in an aquatic plant at different temperatures was been carried out. The results are shown in the table below:

Temperature (°C)	Volume of oxygen released (cm^3)	Time (min)	Rate of photosynthesis ($cm^3\ min^{-1}$)
20	100	5	20
30	200	5	
40	150	5	
50	50	5	
60	0	5	

  a Calculate the rate of photosynthesis for each temperature.
  b Identify the optimum temperature for photosynthesis in this plant.
  c i Explain why the volume of oxygen increased from 20°C to 30°C.
    ii Explain why the volume of oxygen decreased from 30°C to 60°C.
  d Suggest how the researchers could make their data more reliable.

Homeostasis and response

Homeostasis

Homeostasis keeps the internal conditions of the body constant, whatever the outside environment may be.

SNAPIT!

Copy out this table and take a photo of it to get a good overview of all three mechanisms.

The role of homeostasis

The body works best under optimum conditions. Enzymes work best at a temperature of 37°C, and our cells work best with optimum blood glucose levels and water levels. The process that regulates these conditions is called homeostasis. If the internal or external environment changes, it is the role of homeostasis to bring the conditions back to the optimum again.

Examples of homeostasis

	Action if it increases	Action if it decreases
Temperature	Mechanisms to cool the body down	Mechanisms to warm the body up
Blood sugar levels	Release of **insulin** so that cells take up more glucose	Release of **glucagon** to break down glycogen into glucose
Water levels	Less **ADH** is released and more urine is produced	More **ADH** is released and less urine is produced

These will be discussed in more detail on the next few pages.

How homeostasis works

The optimal conditions are under automatic control. If conditions change, nervous or chemical responses bring them back to normal.

Control systems

Receptors →	**Coordination centre** →	**Effector**
Cells that detect stimuli	Receives and processes information from receptors	Responds to restore optimum levels

A stimulus can be any change in the internal or external environment. It is detected by a receptor, that sends a nerve impulse to the coordination centre, usually the brain (but could be the spinal cord or pancreas). The coordination centre sends a nerve impulse to an effector (muscle or gland) which brings about a response.

1 What is homeostasis?

2 Give an example of a condition that needs to be kept at an optimum level in the body.

3 Describe the role of the coordination centre in controlling homeostasis.

The human nervous system

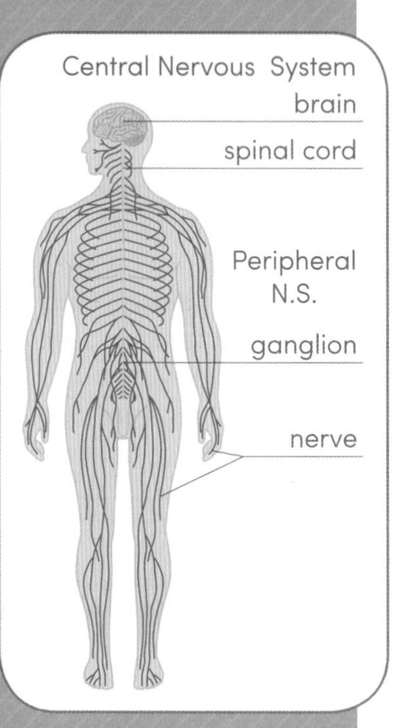

The human nervous system is made of nerve cells, or neurones.

The structure of the human nervous system

The human nervous system is divided into two sections: the central nervous system (CNS) and the peripheral nervous system (PNS).

The central nervous system is made up of the brain and spinal cord and coordinates all of the nerve impulses. The peripheral nervous system (PNS) is made up all of the other nerves and sends and receives nerve impulses.

How the nervous system works

The nervous system allows the body to react to the surroundings and coordinate the body's behaviour. Receptors detect a stimulus and send electrical impulses along the nerves to the spinal cord. The spinal cord sends the electrical impulses to the brain. The brain is a coordinator and sends more electrical impulses back down the spinal cord, and along the nerves to muscles or glands (effectors). The effectors can then respond to the stimulus by either moving a muscle or causing a gland to secrete.

stimulus → receptor → coordinator → effector → response

DO IT!

Write each of the steps in the flow chart on cards and practice putting them into the right order.

WORK IT!

A car driver sees that the traffic light has turned to red. Describe the action of the nervous system. (4 marks)

Retinal cells in the eye are receptors to the red colour (the stimulus). (1)

An electrical impulse is sent to the brain (the coordinator). (1)

The brain sends an electrical impulse to the effectors, the muscles in the foot. (1)

The foot muscles move and put on the brake. This is the response. (1)

NAIL IT!

In Q3 think about the function of each part of the CNS and PNS.

CHECK IT!

1 What are the brain and spinal cord made from?

2 Describe the role of the coordinator.

3 Compare the structure and function of the central nervous system and the peripheral nervous system.

Reflexes

Reflexes are automatic, rapid responses to stimuli that do not involve the conscious part of the brain.

The structure of neurones

There are three types of neurone in a reflex arc; a sensory neurone, a relay neurone and a motor neurone.

The electrical impulses move along the long axon in the direction of the arrows. The dendrites at the end of the neurone pass the electrical impulse to the next neurone by causing a chemical to diffuse across the small gap (the synapse). When the chemical reaches the next neurone, it causes the electrical impulse to continue. The round cell bodies, which contain the nucleus, are in different positions on the different neurones.

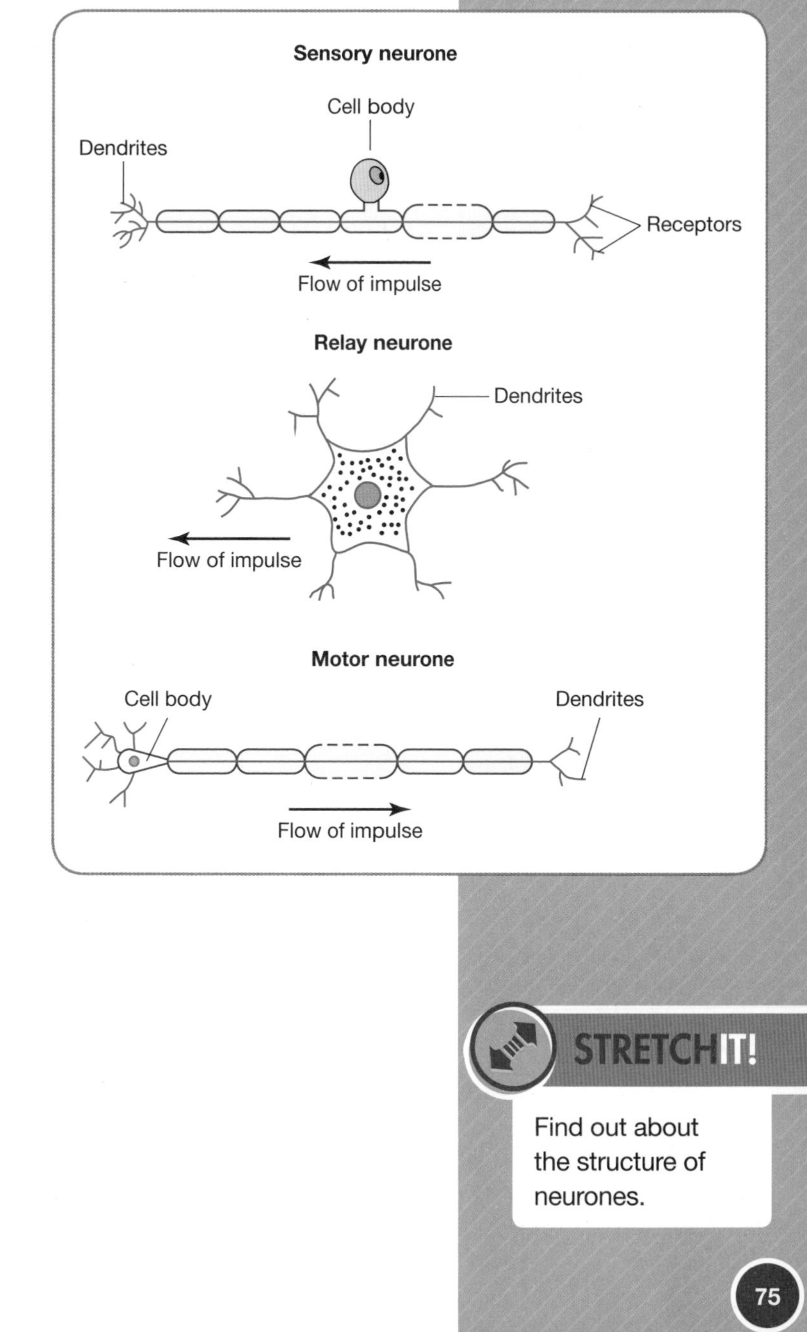

The reflex arc

In a reflex arc, the receptor detects a stimulus and sends an electrical impulse along the sensory neurone to the spinal cord. It passes to a relay neurone in the spinal cord, which sends the electrical impulse straight back along a motor neurone. The motor neurone sends the electrical impulse to an effector which carries out a response. A reflex action does not involve the conscious part of the brain.

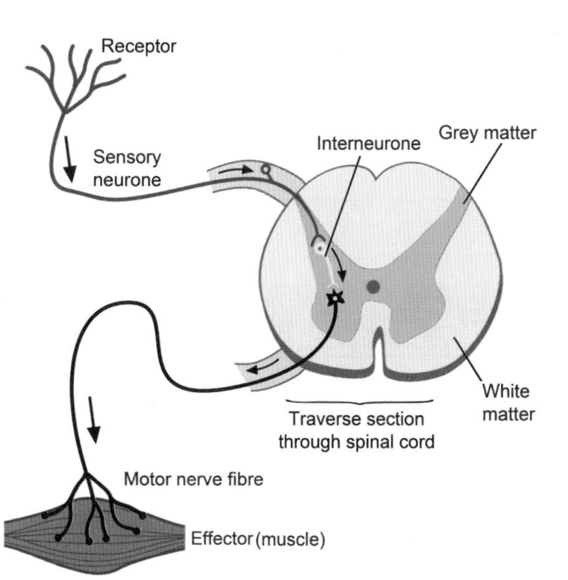

STRETCHIT!

Find out about the structure of neurones.

The importance of the reflex arc

Reflex arcs are quicker than normal reactions and protect us from harm. They allow us to move away from danger or prevent something dangerous from happening.
Some examples of reflexes are:

- blinking when something is near to your eye
- the pupil in your eye contracting in bright light
- coughing reflex when something irritates the top of your windpipe (trachea)
- knee jerk when your lower leg is struck with a small hammer just below your knee.

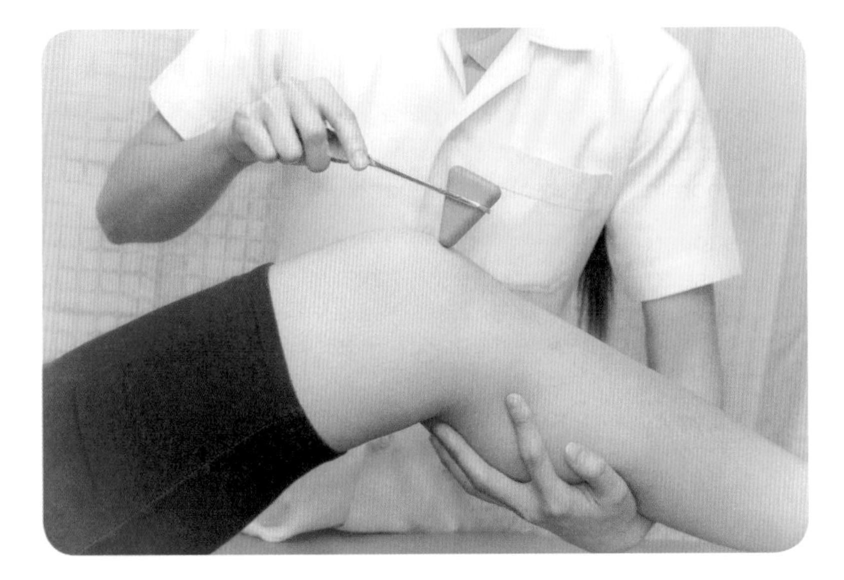

WORKIT!

Compare and contrast the three neurones used in a reflex arc. (4 marks)

Sensory neurones, motor neurones and relay neurones all have an axon and carry an electrical impulse. (1)

Sensory neurones carry an electrical impulse to the spinal cord, motor neurones carry the electrical impulse away from the spinal cord. (1)

Sensory neurones, motor neurones and relay neurones all have dendrites and pass the electrical impulse onto nearby neurones. (1)

Sensory neurones, motor neurones and relay neurones all have a cell body but in different locations on the neurone. (1)

CHECKIT!

1 What is a reflex?

2 Give two examples of a reflex.

3 Describe what happens when a light is shone into someone's eye.

Required practical 7: Investigating the effect of a factor on human reaction time

In this experiment you will be expected to plan an investigation choosing appropriate apparatus and techniques to measure the process of reaction time.

Practical Skills

There are many factors affecting reaction time which you can investigate, for example: before and after drinking a caffeinated drink, right hand versus left hand, morning versus afternoon, and before and after eating.

Popular reaction experiments include timing how long it takes to catch a ruler or to use a computer program where the a button is pressed when an image comes up on the screen.

Make sure that you:

- carry out your reaction experiment before and after your condition/treatment
- repeat your results three times so you can calculate a mean
- show your results as a table and a graph.

DO IT!

Look at your investigation and note down wherever the data was not valid or reliable, and what you could do to improve it.

NAIL IT!

Look at the headings of the columns and the numbers in the table.

WORK IT!

A student wanted to test the effect of caffeine on reaction times. Their data is shown in this table.

Condition	Reaction time 1 (s)	Reaction time 2 (s)	Reaction time 3 (s)	Mean
Before caffeine	0.62	0.65	0.66	0.64
After caffeine	0.34	0.4	0.38	0.37

a Which variables would need to be kept constant during this investigation? (3 marks)

The amount of caffeine drunk. (1) The time of day when the drink was taken. (1) The person who drank the caffeine drink. (1)

b What conclusion can you make about the data? (1 mark)

The reaction time decreases after taking caffeine. (1)

c Correct any mistakes in the table. (2 marks)

The top of the mean column should read, mean time (s). (1) The 0.4 time for after caffeine, reaction time 2 should be 0.40. (1)

CHECK IT!

1 In the Work it! investigation, what is the dependent variable?

2 Why should investigations be repeated?

3 Suggest why drinking caffeine decreases the reaction time.

SNAPIT!

Make a copy of the main regions of the brain with their labels, and take a photo.

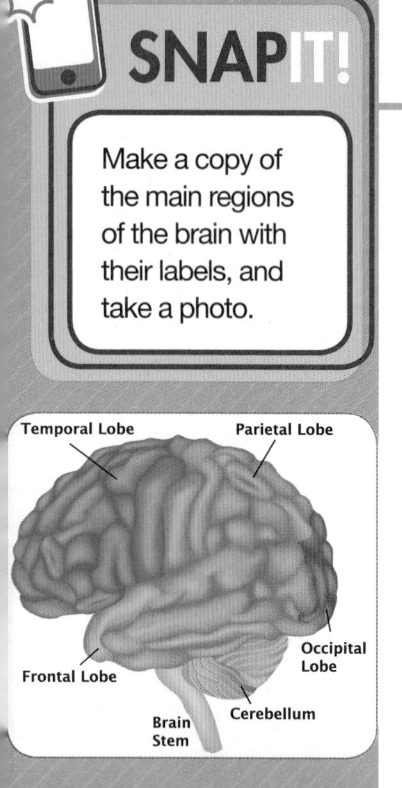

STRETCHIT!

Find out how stem cells are being used in researching Alzheimer's disease.

The brain

The brain controls all of the behaviour in the body.

Regions of the brain

The brain is made of billions of interconnected neurones. Neurones in different regions of the brain carry out different functions. The main regions are shown in the diagram on the left.

The four coloured regions on the figure make up the cerebral cortex. This part of the brain is associated with conscious thought, memory, language and learning. The grey area at the back of the brain is called the cerebellum. This region is responsible for balance and coordinated movements, such as riding a bicycle. The area underneath the brain that leads down to the spinal cord is called the medulla. This area is responsible for coordinating involuntary actions, such as breathing and increasing the heart rate.

Investigating brain function

Most of what we know today about brain function comes from occasions where people have suffered brain damage. It would not be ethical to take out parts of living people's brains to see what happened! Some research has been done on animal brains, but the knowledge is not transferable to humans, as our brains are more complex.

Brain damage and disease

It is difficult to treat brain damage because neurones are replaced very slowly, and the connections between the neurones were formed when a new memory was made. Many people recovering from brain damage have to re-learn skills and recreate these connections.

WORKIT!

Evaluate the benefits and risks of using electrical stimulation and MRI to map the brain. (4 marks)

Electrical stimulation recreates the effect of electrical impulses along neurones. (1) A strong electrical stimulation will damage the brain. (1) MRI is a non-invasive procedure and does not damage the brain. (1) Some people may feel claustrophobic in an MRI machine. (1)

Mapping the brain

Scientists work out the functions of different areas of the brain in two ways:

1. Electrical stimulation – a person's brain is stimulated with a weak electrical current. The subject is asked to describe their experience, e.g. a smell or taste.
2. Magnetic resonance imaging (MRI) – detailed images of the brain are taken to see which parts of the brain are active while the person carries out a task, e.g. listening to music.

CHECKIT!

1. Which area of the brain is associated with learning?
2. What is the function of the cerebellum?
3. Explain why it is difficult to investigate brain function.

The eye

SNAPIT!

Make a copy of the eye diagram and add extra information into the labels about the function of those parts. Take a photo for revising later.

The human eye is a sense organ, containing receptors that detect light and colour.

The structure of the eye

Light enters the eye through the cornea and pupil, and the lens changes shape to focus light onto the retina at the back of the eye. The retina contains two types of light receptor, rods and cones. The cones detect light intensity and colour, and work best in bright light. The rods do not detect colour and work best in low light. The retina creates an upside-down image, and sends this image via an electrical impulse along the optic nerve to the brain.

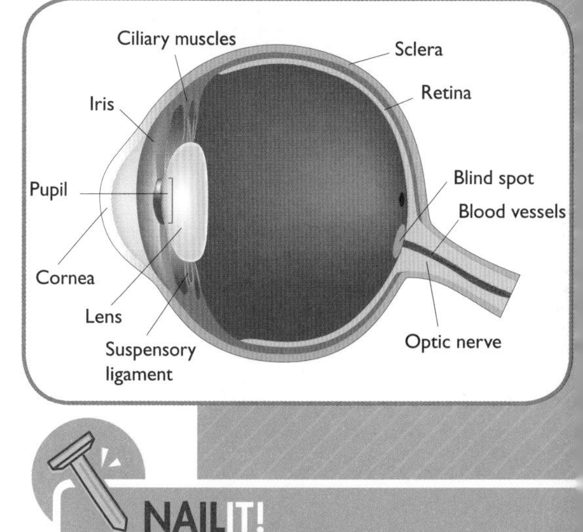

Functions of each part of the eye

Part of the eye	Function
Retina	Covered in light-sensitive receptors that detect light and create an image
Optic nerve	Transmits electrical impulses from the retina to the brain
Sclera	White outer coat that protects the eye
Cornea	Transparent front of the eye that carries out initial focusing
Iris	Controls the amount of light that can enter the eye
Ciliary muscles	Contracts and relaxes to change the shape of the lens
Suspensory ligaments	Connects the ciliary muscles to the lens

NAILIT!

Remember the cornea and the lens refract the light towards the retina.

DOIT!

Put each part of the eye and the function onto revision cards and match them up.

CHECKIT!

1 Which part of the eye changes the shape of the lens?

2 Describe the role of the sclera.

3 Compare and contrast the function of rods and cones of the retina.

WORKIT!

Describe how light is focused onto the retina. (3 marks).

The cornea carries out initial focusing of light (1) The iris controls the amount of light that enters the eye through the pupil. (1) The lens changes shape to focus the light onto the retina. (1)

Focusing the light

SNAPIT!

Make a copy and take a photo of the eye diagrams on this page, and become familiar with how the lenses change shape and how the corrective lenses work.

The lens in the eye changes shape in order to focus the light.

Accommodation of the eye

Accommodation is the process of changing the shape of the lens to focus on near or distant objects.

Near objects

When an object is near, the ciliary muscles contract, and the suspensory ligaments loosen. The lens becomes thicker and refracts light rays strongly, to focus light onto the retina.

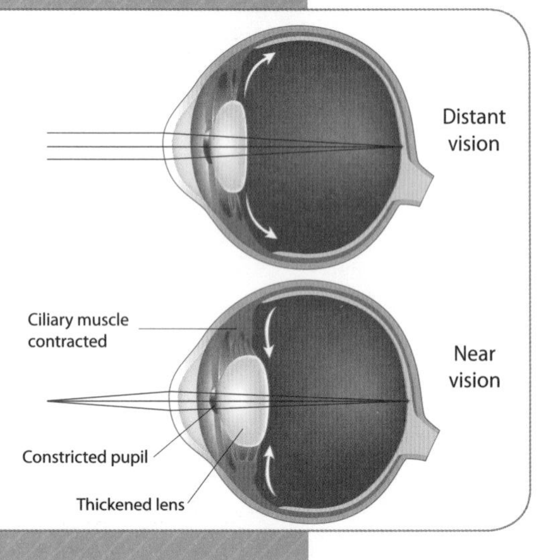

Distant objects

When an object is distant, the ciliary muscles relax, and the suspensory ligaments are pulled tight. The lens becomes thinner and refracts light slightly, to focus light onto the retina.

Use of lenses

When the lens does not focus the light onto the retina properly, this can be corrected by using the correct spectacle lens.

NAILIT!

Myopia (short-sightedness)

This is when distant objects are not focused clearly. The image is focused in front of the retina. It is corrected with a concave lens.

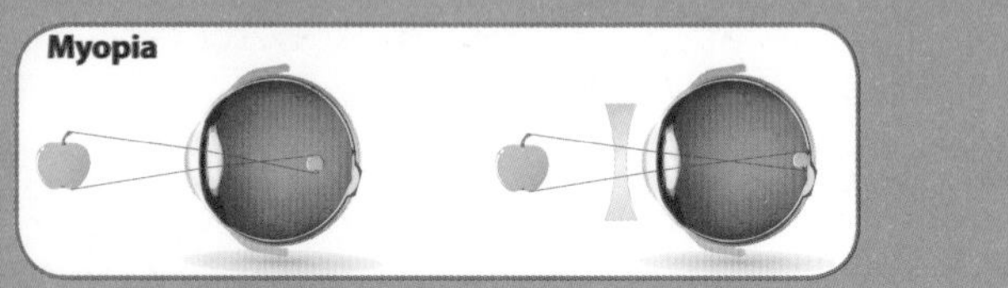

NAILIT!

Hyperopia (long-sightedness)

This is when near objects are not focused clearly. The image is focused behind the retina. It is corrected with a convex lens.

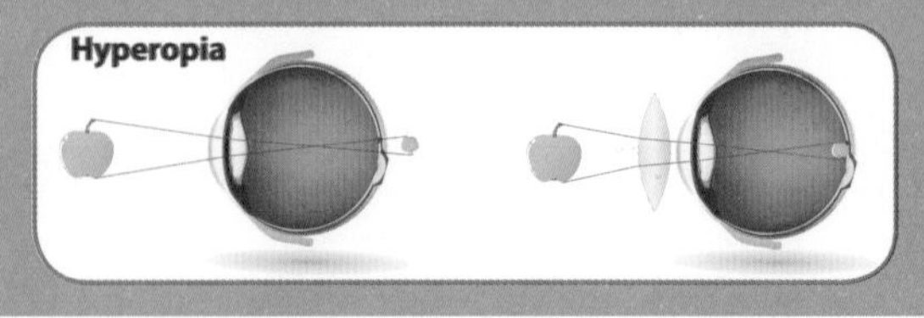

CHECKIT!

1. Explain how lenses are used to correct hyperopia (long-sightedness).
2. Describe the role of the ciliary muscles in the accommodation of the eye.

Control of body temperature

The body temperature in humans is controlled by homeostasis.

Body temperature monitoring

The optimum temperature for the body is 37°C, but the normal range for body temperature is between 36.1 – 37.2°C.

- If the temperature is too low, (below 35°C), then a person has hypothermia
- If the temperature is too high (above 38°C), then a person may have a fever or hyperthermia.

The body temperature is monitored and controlled by the thermoregulatory centre in the brain. The thermoregulatory centre contains thermoreceptors (receptors that detect a change in temperature) and monitors the temperature of the blood directly. There are also thermoreceptors in the skin. When these detect a change in temperature, they send electrical impulses to the thermoregulatory centre.

Responses to change in temperature

When the thermoreceptors detect a change in temperature, the thermoregulatory centre sends electrical impulses to many effectors around the body to increase or decrease the temperature.

Body temperature is too high	Body temperature is too low
Blood vessels close to the skin dilate (vasodilation) to release more heat energy to the environment.	Blood vessels close to the skin constrict (vasoconstriction) so that less heat energy is released to the environment.
Sweat is produced by sweat glands. As the sweat evaporates, it takes some heat energy into the environment.	Skeletal muscles contract rapidly (shivering). This generates some heat energy.

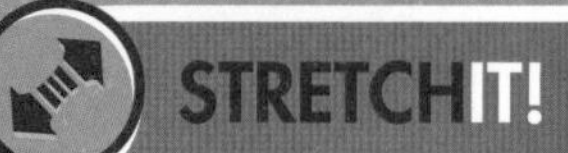

Find out how hypothermia and hyperthermia affect the body.

DOIT!

Write the responses to temperature change onto revision cards and match them to when temperature is 'too high' and 'too low'.

WORKIT!

A person has a body temperature of 35.5°C. Describe and explain how the body returns the body temperature back to 37°C. (4 marks)

The blood vessels near to the skin in the body constrict (vasoconstriction) (1) to reduce the amount of heat energy released into the environment. (1) The muscles cause shivering (1) to generate more heat energy. (1)

CHECKIT!

1 What could you say about a person who has a body temperature of 39°C?

2 How does sweating cool a person down?

3 Describe how temperature is monitored by the body.

Give examples of what the effectors in the body do and explain how this changes body temperature.

Human endocrine system

SNAP IT!

Make a copy of the endocrine system diagram and learn the labels for the glands.

DO IT!

Learn the names of all of these glands and at least one example of a hormone that they release.

WORK IT!

Describe the role of the 'master gland'. (4 marks)

The pituitary gland (1) releases hormones in response to the environment. (1) The hormones produced by the pituitary gland target another endocrine gland (1) and cause them to release a hormone. (1)

The human endocrine system is made up of many glands that secrete hormones into the blood. Hormones act as chemical messengers.

How the endocrine system works

Each gland makes one or more hormones, which are then secreted directly into the bloodstream.

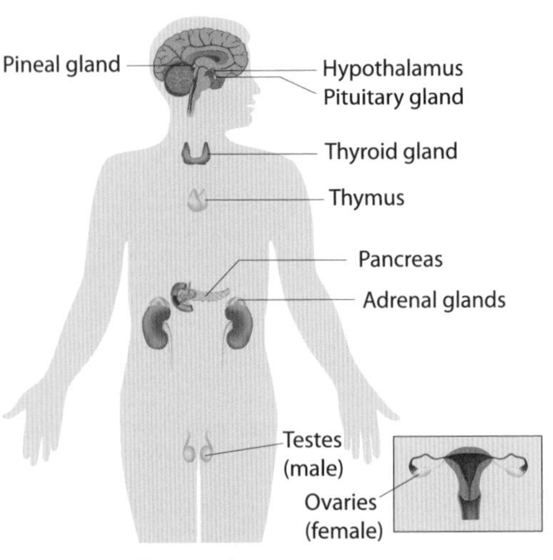

The Endocrine System

Gland	Hormone
Pituitary gland	Many hormones, including growth hormone, follicle-stimulating hormone (FSH) and luteinising hormone (LH)
Pancreas	Insulin and glucagon
Thyroid	Thyroxine
Adrenal gland	Adrenaline
Ovary	Oestrogen and progesterone
Testes	Testosterone

The blood carries the hormone to a target organ or cells, where the hormone has an effect.

The pituitary gland

The pituitary gland in the brain controls many of the other glands in the body by releasing hormones that affect them. For example, the pituitary releases thyroid-stimulating hormone (TSH), which targets the thyroid and causes it to release thyroxine. For this reason, the pituitary gland is often called the 'master gland'.

Endocrine system versus nervous system

The endocrine system acts more slowly than the nervous system, but the effects of the endocrine system last for a longer time.

CHECK IT!

1. What is the name of the endocrine gland on top of the kidneys?
2. Give one example of an endocrine gland and the hormone that it produces.
3. State one similarity and one difference between the endocrine and nervous systems.

Control of blood glucose concentration

STRETCH IT!

Find out how a lack of insulin affects people with diabetes.

SNAP IT!

Make a copy of the diagram of the negative feedback cycle and take a photo.

Blood glucose concentration is monitored and controlled by the pancreas.

The role of the pancreas

The pancreas monitors the blood glucose concentration as the blood passes through it.

- If the blood glucose concentration is too high, for example, just after a meal, then the pancreas releases insulin into the blood.
- If the blood glucose concentration is too low, for example, several hours after eating, then the pancreas releases glucagon into the blood.

Insulin

Insulin is a hormone that targets all cells in the body. It causes the cells to take up glucose (a type of sugar). This decreases the concentration of glucose in the blood back to the normal levels. In the liver and muscle cells, any excess glucose is converted into glycogen for energy storage.

Glucagon

Glucagon is a hormone that targets the liver and muscle cells. Glucagon causes the liver and muscle cells to break down glycogen into glucose and release it into the blood. This increases the concentration of blood glucose back to the normal levels.

Negative feedback cycle

The actions of insulin and glucagon can be described as a negative feedback cycle. In a negative feedback cycle, when something increases or decreases from the normal level, the effect works in the opposite direction to bring it back to the normal level. The diagram on the right shows the negative feedback cycle of insulin and glucagon.

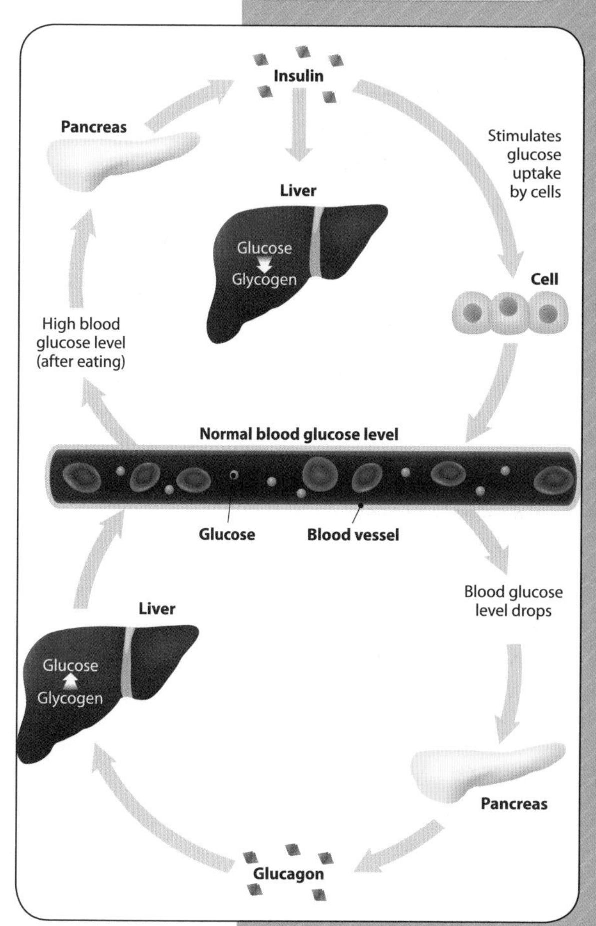

CHECK IT!

1 Where is glucagon made?

2 Describe the role of glycogen.

H 3 Explain how the blood glucose level is maintained using a negative feedback cycle.

Diabetes

In people with diabetes, either no insulin is produced, or the body stops responding to the insulin in the blood.

Make a table to compare type 1 and type 2 diabetes.

Types of diabetes

Not being able to control body sugar levels is dangerous for the body. The normal blood glucose level is 90 mg/dl, rising to 135 mg/dl immediately after a meal.

- When the blood glucose level is too low, it is called hypoglycaemia. This can cause shaking, dizziness, and even coma.
- When the blood glucose level is too high, it is called hyperglycaemia. This can cause kidney damage or damage to the eyes.

Type 1 diabetes

Type 1 diabetes is caused when the pancreas does not produce enough insulin. It is often identified in children, and is thought to have a genetic cause. People with type 1 diabetes have to inject insulin several times a day.

Type 2 diabetes

Type 2 diabetes is caused by lifestyle factors, such as obesity and poor diet. The pancreas still produces insulin but the body does not respond to it in the correct way. People with type 2 diabetes have to follow a carbohydrate-controlled diet and exercise regularly to control their condition.

Blood glucose levels in diabetes

After a meal, the blood glucose levels increase in people with and without diabetes. In those without diabetes, insulin is released and the blood glucose levels start to return to normal. For those with diabetes, the blood glucose levels remain high unless treated.

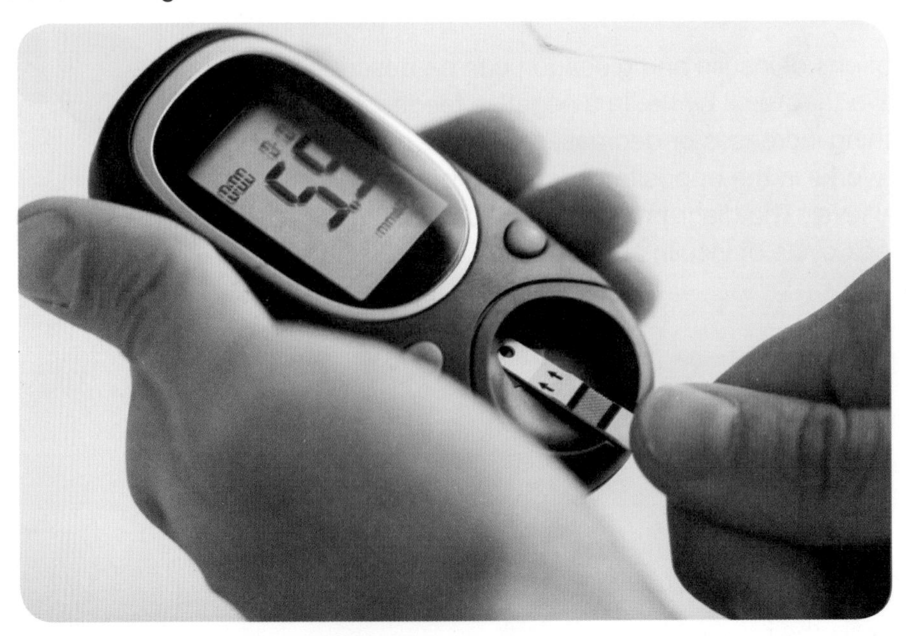

Blood glucose monitor

WORKIT!

This graph shows the blood glucose levels in a person with type 2 diabetes and a person who does not have diabetes. Describe and explain the line of the graph for the person with type 2 diabetes.
(4 marks)

Blood glicose conc. (mg/dl)
200
180
160
140
120
100
80
type 2 diabetic
normal
1 2 3 4 5
Hours

SNAPIT!

Take a photo of this graph and check that you understand what it means.

The blood glucose level starts at 135 mg/dl and increases to 200 mg/dl in 1.25 hours. (1)

From 1.5 to 5.5 hours the blood glucose levels decrease slowly to 150 mg/dl. (1)

This is because body does not respond in the correct way to the insulin produced by the pancreas, (1) and glucose in the blood is not taken up by the cells. (1)

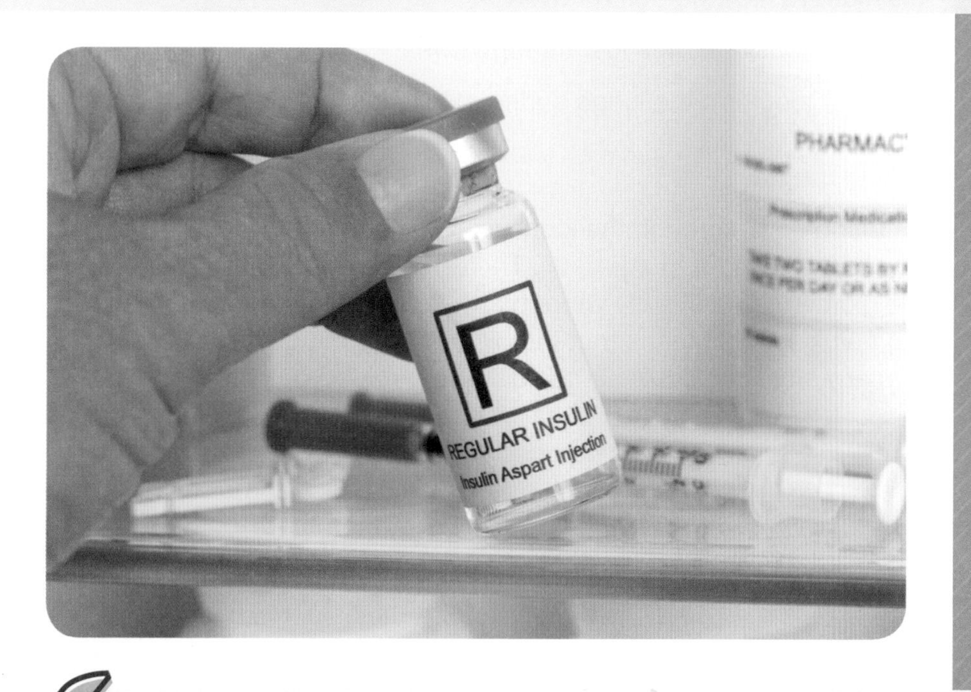

CHECKIT!

1. What treatment is given to people with type 1 diabetes?
2. Describe the symptoms of a person with diabetes when the blood glucose levels are too low.
3. What advice would you give to a person who is worried about developing type 2 diabetes?

Maintaining water and nitrogen balance in the body

NAILIT!

You will not be expected to know the structure of the rest of the kidney, nephron or bladder.

SNAPIT!

Make a copy of the diagram of the glomerulus and label what is happening in the glomerulus and Bowman's capsule. Take a photo.

For body cells to function correctly, water and nitrogen levels in the body must be regulated.

Osmotic changes

Osmosis is the diffusion of water from a more dilute solution to a more concentrated solution (see page 23).

- If the blood or the tissue fluid surrounding the cells is too concentrated, water will move out of the cells. The cells could become shrivelled or crenated and unable to function normally.
- If the blood or the tissue fluid surrounding the cells is too dilute, water will move into the cells. The cells could burst and would no longer function.

The levels of water and nitrogen ions are controlled by the kidneys. However, the kidneys have no control over:

- water, ions and urea lost through sweat
- water lost from the lungs during exhalation.

The kidneys

As blood flows through the kidneys, it is filtered through a ball of blood capillaries called a glomerulus. The arteriole (small artery) taking the blood to the glomerulus is slightly wider than the arteriole taking blood away from the glomerulus. This means that the blood inside the glomerulus is under pressure and any small molecules in the blood are forced out into a part of the kidney called the Bowman's capsule. This is called ultrafiltration. The water, ions and urea flow from the Bowman's capsule into the nephron, where they move into the bladder to be removed from the body as urine.

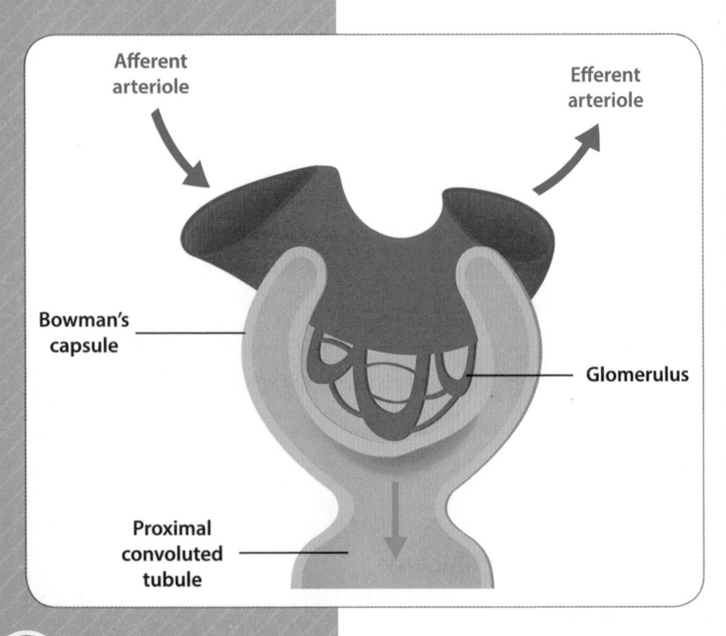

Small molecules, such as glucose, are also removed by ultrafiltration, but these are selectively reabsorbed back into the blood as they pass along the nephron. Some ions and water that are needed by the body are also reabsorbed. Large molecules, such as protein, are too large to pass out of the blood capillaries and remain in the blood.

Before and after filtration

	In blood before filtration	In blood after filtration	In nephron after filtration
Water	Depends on the water balance of body		
Excess ions	High	Low	High
Urea	High	Low	High
Glucose	High	Reabsorbed – high	None
Proteins	High	High	None

Digestion of proteins

Unneeded proteins in the body are digested, resulting in an excess of amino acids. These are sent to the liver to be broken down (deaminated) into ammonia. Ammonia is toxic, so it is converted into urea, which travels in the blood to the kidneys to be excreted in urine.

STRETCHIT!

Find out what happens to the body when the kidneys do not function properly.

WORKIT!

The graph shows the concentration of glucose, ions and urea in the blood before filtration, and after filtration in the nephron. Describe and explain this data. (6 marks)

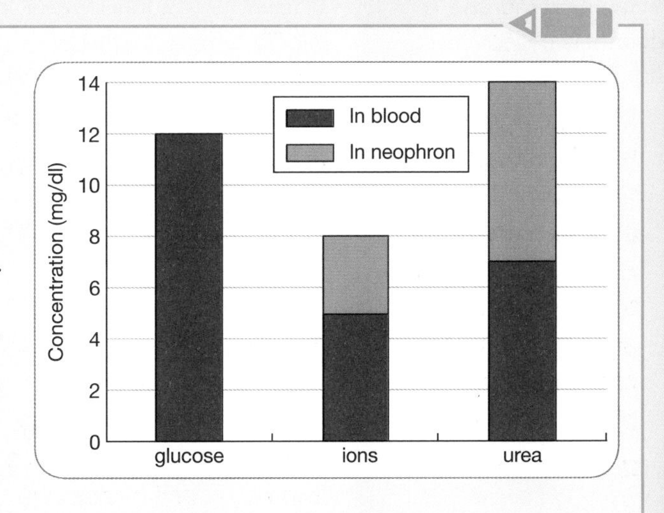

There are 12 mg/dl of glucose in the blood before filtration and no glucose in the nephron after filtration. (1) This is because all of the glucose is selectively reabsorbed from the nephron back into the blood. (1)

There are 5 mg/dl of ions in the blood before filtration and 3 mg/dl of ions in the nephron after filtration. (1) This is because 2 mg/dl of ions were reabsorbed back into the blood. (1)

There were 5 mg/dl of urea in the blood before filtration and 5 mg/dl of urea in the nephron after filtration. (1) This is because all of the urea was removed from the body in the urine. (1)

1 Name two substances that are removed from the blood by filtration in the kidneys.

2 Describe the effect on cells if the surrounding tissue fluid has too many ions.

3 Suggest what would happen to the concentration of urine if there was too much water in the body.

ADH

Anti-diuretic hormone (ADH) is a hormone that acts on the kidneys to control the amount of water lost from the body.

DO IT!

Make a flow diagram of the steps involved in releasing ADH and its effects.

The effect of ADH

The hypothalamus in the brain monitors and maintains the water levels in the body.

- If there is too little water, the blood is too concentrated and the hypothalamus causes the pituitary gland to release ADH.
- If there is too much water, the blood is too dilute and the hypothalamus causes the pituitary gland to stop releasing ADH.

This is a type of negative feedback.

STRETCH IT!

Find out how ADH acts on membrane channels called aquaporins.

Too little water

If there is too little water in the body, the pituitary gland secretes ADH into the blood where it travels to the kidneys. In the kidneys, ADH causes more water to be reabsorbed back into the blood. This happens in a part of the kidney tubules called the collecting duct. In the presence of ADH, the collecting duct is more permeable to water so more water leaves the collecting duct, and moves back into the blood. If more water is reabsorbed back into the blood, the urine produced in the collecting duct is more concentrated.

Too much water

If there is too much water in the body, the pituitary gland does not secrete ADH. In the kidneys, the absence of ADH causes less water to be reabsorbed back into the blood. If less water is reabsorbed back into the blood, the urine produced in the collecting duct is more dilute.

WORK IT!

A person who drinks a lot of water throughout the day observes that their urine is very pale. Explain why that is. (4 marks)

The hypothalamus detects that the blood concentration is too dilute, (1) so no ADH is released from the pituitary gland. (1) Less water is reabsorbed in the collecting duct of the kidneys, (1) so the urine is more dilute. (1)

NAIL IT!

Look back to pages 83 and 94 for more on negative feedback cycles.

CHECK IT!

H1 What is the role of ADH?

H2 Describe what happens in the kidney if the blood is too concentrated.

H3 Explain how the release of ADH is a form of negative feedback.

Dialysis

If a person has kidney disease, they may need dialysis.

The purpose of dialysis

If a person's kidneys cannot filter their blood then they will become very ill. The only way to treat kidney disease of this kind is to have a kidney transplant or to use a dialysis machine. The dialysis machine filters the blood and removes excess ions, excess water and urea.

The process of dialysis

How dialysis works

The person who needs treatment has tubes attached to a needle in their arm. The tubes take their blood through the dialysis machine and then back into their body. It can take up to five hours to filter the blood, and this will need to be done several times a week. The blood moves through the dialysis machine, where it is separated from the dialysate (a fluid containing ions and sugars in the same concentration as the blood) by a partially permeable membrane.

The concentration of urea is higher in the blood compared to the dialysate, so this moves into the dialysis fluid down a concentration gradient by diffusion. If the ion concentration in the blood is higher than normal, any excess ions also move into the dialysate.

The advantages and disadvantages of dialysis and transplants

	Advantages	Disadvantages
Dialysis	No need for surgery Available (no waiting list)	Time consuming; risk of infection; patient must limit protein and salt in diet
Kidney transplant	No need for dialysis	Shortage of organ donors; risk of rejection of kidney; need to take immunosuppressant drugs

NAILIT!

Think about the concentration of the ions in the dialysate compared to the blood.

WORKIT!

Explain why a person who needs dialysis should be careful about the concentration of salt in their diet. (3 marks)

Any excess salt ions need to be removed from the blood by the dialysis machine. (1) If the salt ion concentration in blood is higher than the concentration in dialysate (1) then the excess salt ions will move into the dialysate by diffusion. (1)

CHECKIT!

1 What is the purpose of a dialysis machine?

2 Name two advantages of using dialysis rather than having a kidney transplant.

3 Describe how urea is removed from the blood by the dialysis machine.

Hormones in human reproduction

Human reproduction involves the interaction of many hormones.

Hormones during puberty

During puberty, the sex hormones are released in males and females. These hormones cause secondary sex characteristics to develop.

- In males testes secrete testosterone. Males develop deeper voices, chest and face hair, a more muscular body, and pubic hair. The genitals develop and sperm are produced.
- In females ovaries secrete oestrogen and progesterone. Females develop breasts, broader hips and pubic hair. Menstruation starts.

SNAP IT!

Make a sketch of the graph and check that you understand what it means.

The menstrual cycle

Menstruation, or the menstrual cycle, lasts an average of 28 days, and is controlled by four hormones:

- Follicle stimulating hormone (FSH) – causes eggs (ova) in the ovaries to mature.
- Luteinising hormone (LH) – stimulates the release of an ovum (ovulation).
- Oestrogen – repairs the lining of the uterus after menstruation.
- Progesterone – maintains the uterus lining.

Make a flow diagram of the steps involved in releasing the different hormones during the menstrual cycle.

HORMONE LEVEL
LH
FSH
ESTROGEN
PROGESTERONE

FOLLICULAR DEVELOPMENT

DAY OF CYCLE
1. 2. 3. 4. 5. 6. 7. 8. 9. 10. 11. 12. 13. 14. 15. 16. 17. 18. 19. 20. 21. 22. 23. 24. 25. 26. 27. 28.

FOLLICULAR PHASE
OVULATORY PHASE
LUTEAL PHASE

H

Interaction of hormones in the menstrual cycle

The pituitary gland releases FSH at the start of the menstrual cycle. FSH acts on the ovaries and causes many ova to mature. It also causes the ovaries to secrete oestrogen. Oestrogen slows down the production of FSH, so that usually only one ovum reaches full maturity. Oestrogen also stimulates the pituitary gland to release LH.

LH peaks around day 14 of the menstrual cycle, causing the mature ovum to be released from one of the ovaries. The release of the ovum causes large amounts of progesterone to be secreted by the ovaries. This causes a negative feedback loop, which stops FSH and LH from being produced by the pituitary gland. This in turn causes a decrease in oestrogen.

WORKIT!

Describe what happens to the hormone levels in a female body if an ovum is not fertilised. (3 marks)

The progesterone levels decrease. (1)

The oestrogen levels decrease. (1)

This means that FSH is no longer inhibited and the levels of FSH increase. (1)

NAILIT!

To answer this question, look at day 27 on the menstrual cycle graph on the previous page.

STRETCHIT!

At Higher Tier you should be able to explain the interactions of FSH, oestrogen, LH and progesterone, and their role in the control of the menstrual cycle.

1. Name the hormone involved in developing male secondary sex characteristics.
2. What is the role of LH?
3. Suggest what happens to the levels of hormone in the female body if the ovum is fertilised.

Contraception

Contraception is any method used to prevent pregnancy.

Non-hormonal methods of contraception

This type of contraception uses barrier methods to prevent sperm reaching the egg. Non-hormonal methods include:

- using condoms/diaphragms – trap sperm
- spermicidal agents – kill or disable sperm
- intrauterine device which prevents the implantation of an embryo (may also release hormones)
- surgical methods (sterilisation) – cut and tie sperm ducts in males to prevent sperm leaving the penis, or cut and tie oviducts (Fallopian tubes) in females to prevent eggs travelling from the ovaries to the uterus
- abstaining from intercourse around the time that an egg may be in the oviducts.

Hormonal methods of contraception

Some methods of contraception use hormones to prevent the release of an egg during the menstrual cycle. For example:

- oral contraceptives (the pill) – contain oestrogen and progesterone so that FSH is inhibited, and no eggs mature (see pages 90–91).
- injection, skin patch, or implant under the skin – contain progesterone to inhibit the maturation and release of eggs. These last several months.

DOIT!

Write each method of contraception onto revision cards.

STRETCHIT!

Find out the side effects of using hormonal methods of contraception.

WORKIT!

Evaluate the hormonal and non-hormonal methods of contraception. (4 marks)

Condoms do not have any side effects, but are not 100% effective at preventing pregnancy. (1)

Oral contraceptives are effective at preventing pregnancy, but have some side effects on the female body. (1)

Condoms and oral contraceptives allow couples to choose the time tov start a family. (1)

Surgical methods are very effective at preventing pregnancy but are difficult to reverse. (1)

CHECKIT!

1 Give an example of a non-hormonal method of contraception.

2 Describe how the hormones in oral contraceptives prevent eggs from maturing.

3 Suggest why some women may prefer to use a hormonal injection, rather than use oral contraceptives.

H

Using hormones to treat infertility

FSH and LH are used in in vitro fertilisation as fertility drugs.

In vitro fertilisation (IVF)

IVF is a type of fertilisation that happens outside of the body. Eggs are surgically removed from the mother and fertilised by the father's sperm in a laboratory. The fertilised eggs develop into embryos and are then placed surgically into the woman's uterus.

FSH and LH are given to the woman in order to produce enough eggs for IVF. These hormones cause several eggs to reach maturity so that several eggs can be fertilised by IVF at the same time.

Advantages and disadvantages of IVF

Advantages

- It allows the mother to give birth to her own baby.
- The baby will be the genetic offspring of the mother and father.

Disadvantages

- The success rates are not high – many couples need to have several rounds of IVF before a successful birth.
- It can lead to multiple births (two, three or more babies at once) as several embryos are implanted at once. This is less safe for the mother and the babies, and can lead to health problems.
- It is emotionally and physically stressful.

STRETCH IT!

Find out about mothers who have had multiple births through IVF.

DO IT!

Write each advantage and disadvantage of IVF onto revision cards.

NAIL IT!

Think about techniques and equipment that would be needed.

WORK IT!

IVF has been around since 1978. Suggest why IVF was not available before this date. (3 marks)

Developments in light microscopes that allowed the sperm and eggs to be viewed. (1) Developments in surgical techniques to remove eggs and implant embryos safely. (1) Developments in forms of FSH and LH that can be safely given to women undergoing IVF. (1)

CHECK IT!

H1 Which hormones are given to women as fertility drugs?

H2 Describe what happens during IVF.

H3 What social and ethical issues are associated with IVF treatments?

Negative feedback

Negative feedback is a mechanism that keeps the body functioning at set levels. If something goes above or below the set level, negative feedback brings it back again.

Adrenaline

Adrenaline is a hormone produced by the adrenal gland. It is secreted into the bloodstream when the body is under stress, for example, if a person is scared or excited. Adrenaline has several effects that help prepare the body for 'fight or flight'.

Heart rate increases – to increase blood flow around the body; breathing rate increases – to increase the amount of oxygen in the blood; blood is redirected to the muscles – more blood flows to the muscles and less to the skin and intestines; stimulates the liver – to breakdown glycogen into glucose to provide more glucose for muscles to use for respiration; dilated pupils – more light enters the eyes.

Thyroxine

Thyroxine is a hormone secreted by the thyroid gland. Thyroxine stimulates the basal metabolic rate, which is important in growth and development.

If the thyroxine levels in the body are low, the thyroid gland is stimulated by the hypothalamus to secrete more; If the thyroxine levels in the body are high, the thyroid gland is stimulated by the hypothalamus to stop producing thyroxine.

This is a type of negative feedback.

DO IT!

Draw a diagram showing the negative feedback cycle for thyroxine.

WORK IT!

A person is feeling very tired and lacks in energy. Describe what will happen inside the body to return the person to their normal basal metabolic rate. (3 marks)

The hypothalamus stimulates the thyroid gland. (1) The thyroid gland secretes thyroxine. (1) Thyroxine stimulates the basal metabolic rate. (1)

CHECK IT!

H 1 Which organ system is the thyroid gland a part of?

H 2 Name two effects of adrenaline.

H 3 Explain why the secretion of adrenaline is not controlled by negative feedback.

Plant hormones

Plants use hormones to coordinate and control growth, and respond to their environment.

The role of hormones in plant responses

Plant responses to the environment are controlled by the action of plant hormones, such as auxin. Auxin is produced in the shoot tips and root tips of a plant. There are several important plant responses.

- Phototropism – plants responding to light.
- Geotropism – plants responding to gravity.

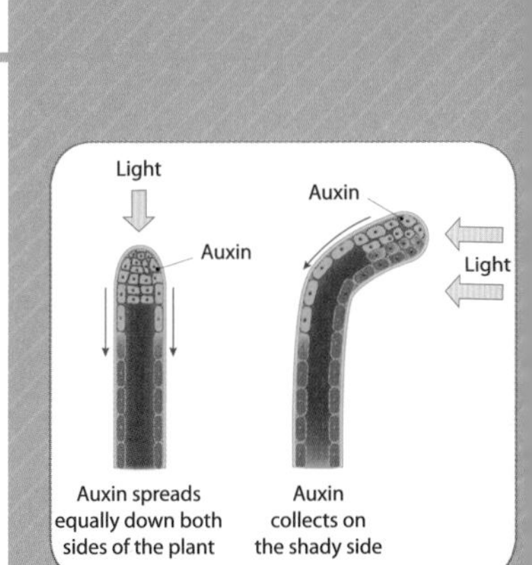

Phototropism and auxin

Auxin causes the cells in the shoot to elongate (get longer). More auxin gathers on the side of the shoot that is in the shade, which causes the shoot tip to bend in the direction of the light.

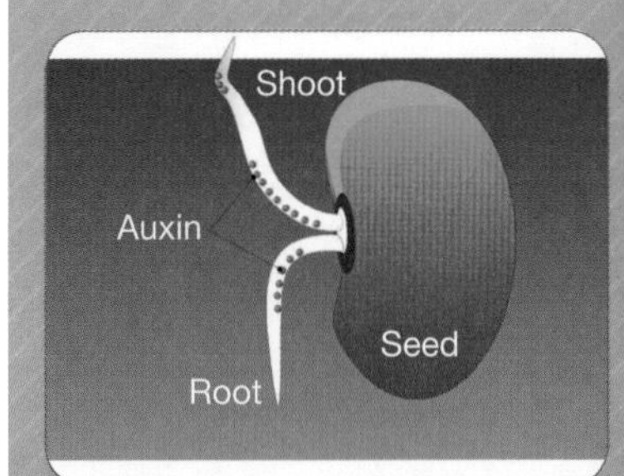

Geotropism and auxin

In the root tips, auxin inhibits elongation. More auxin gathers on the side of the root that is facing downwards. This causes the root tip to bend downwards.

In the shoot tips, auxin promotes elongation. More auxin gathers on the side of the shoot that is facing downwards, causing the shoot tip to bend upwards.

Uses of plant hormones

Auxins, gibberellins and ethene are all used in farming and gardening to control plant growth.

H

- Gibberellins – initiate seed germination. They are also used to promote flowering and increase fruit size.
- Ethene – controls cell division and ripening of fruits. This is used by the food industry to control ripening during storage and transport.
- Auxins are used as weed killers and rooting powders. In tissue culture (used to clone new plants), auxin is used to promote growth.

NAILIT!

You will not be expected to know the mechanisms of how gibberellins and ethene work.

WORKIT!

Compare and contrast the action of auxin in the root tip and the shoot tip of a plant. (4 marks)

Auxin gathers on the lower part of the root and shoot tip. (1) In the root tip, auxin inhibits elongation, (1) and in the shoot tip auxin promotes elongation. (1) The root tip bends downwards, and the shoot tip bends upwards,. (1)

CHECKIT!

1. What is phototropism?
2. Name the plant hormone that controls cell division.
3. Compare the action of auxin in the shoot tip during phototropism to the action of auxin in the shoot tip during geotropism.

Required practical 8: Investigating the effect of light or gravity on the growth of newly germinated seedlings

In this practical you will need to record results both as length measurements and as accurate, labeled biological drawings to show the effects. You will be expected to use your knowledge of plant responses to explain your observations.

Practical Skills

The effect of gravity

The most common method of investigating the effect of gravity on seedlings is to alter the position of the seedlings so that a different part of them is facing downwards.

Practical Skills

The effect of light

A common method for investigating the effect of light on seedlings is to place a light source on one side of the seedlings and observe the growth of the shoots. Then, move the light source and observe the growth of the shoot again.

WORKIT!

A newly germinated seedling has been rotated on a daily basis. The length and direction of the root growth is shown in the table

Time (days)	Length of root (cm)	Direction of root growth
1	2.5	Down
2	5.1	Down
3	7.6	Down
4	9.2	Down

Use your knowledge of geotropism to explain this data. (4 marks)

Root tips show positive geotropism/always grow downwards. (1)

Auxin gathers in the lower side of the root tip. (1)

This inhibits elongation in the cells on the lower side of the root tip. (1)

The cells on the upper side of the root tip elongate and bend the root tip towards gravity. (1)

CHECKIT!

1 What is the name of the response of plants to light?

2 Describe the effects of auxin on cells in roots and shoots.

3 Suggest why it is important that plant shoots grow towards the light.

Homeostasis and response

REVIEW IT!

1 a Define homeostasis.

b In thermoregulation, which is the:

i Receptor ii Coordinator iii Effector?

c Draw a negative feedback diagram to show what happens in the control of blood glucose levels.

d Explain what happens in people who do not produce insulin.

2 a Which gland in the endocrine system produces:

i Adrenaline ii Oestrogen iii Glucagon?

b ADH is a hormone that is made by the pituitary gland. Explain how ADH affects the volume of urine that is produced by the kidney.

c In people with limited kidney function, dialysis or a kidney transplant is needed to keep the person healthy.

i Explain how dialysis works.

ii Evaluate the advantages and disadvantages of a kidney transplant.

3 a What part of the brain is responsible for:

i memory ii breathing rate?

b The lens of the eye focuses the light onto the retina at the back of the eye.

i Draw how the light is focused in a person with myopia.

ii What type of lens is needed to correct myopia?

4 a Name two neurones in the nervous system.

b Compare a voluntary nerve response to a reflex.

c Students A and B have carried out an investigation into reaction times before and after drinking a drink containing 50 mg caffeine. Their results are shown below:

Condition	Student A	Student B
Before 50 mg caffeine	0.45	0.68
After 50 mg caffeine	0.40	0.59

i Student A says this proves that caffeine increases reaction rates. Are they correct? Justify your answer.

ii How could they improve their investigation?

Inheritance, variation and evolution

Sexual and asexual reproduction

Sexual reproduction requires two parents, whereas **asexual** reproduction only requires one parent.

Sexual reproduction

Sexual reproduction requires a male and a female. Both the male and the female produce gametes (sperm or pollen in males, eggs or egg cells in females.) Gametes are produced by meiosis (see page 100) and have half of the genetic information of a body cell. During meiosis, the genetic information is mixed, so that each of the gametes has different genetic material. This leads to variety in the offspring.

When the male gamete and the female gamete meet, the nuclei from each fuse together. This makes the first cell of the new organism – the zygote. In plants, the zygote is a seed. In animals, the zygote divides by mitosis into a ball of cells called an embryo. In mitosis, all of the cells are genetically identical.

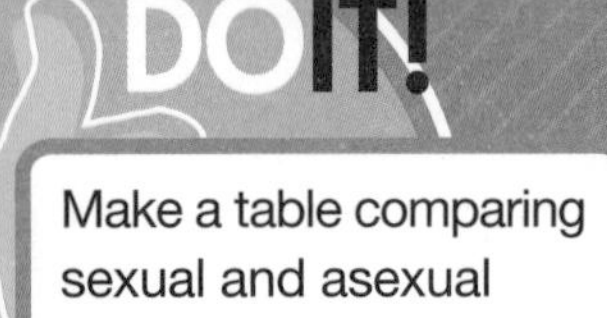

Make a table comparing sexual and asexual reproduction.

Asexual reproduction

Some organisms can reproduce without another parent, using mitosis. All of the offspring are genetically identical to the parent. These are natural clones. There are several different mechanisms of asexual reproduction:

- Binary fission – bacteria divide their genetic material and elongate to double their normal size. They then break into two identical daughter cells.
- Budding – yeast double their genetic material and sub-cellular structures and form small buds on their surface. The buds develop and then break off from the parent yeast.
- Runners, bulbs and tubers – plants use these methods to grow new plants. For example, strawberry plants send out runners. Where they find soil, the runner develops roots and forms a new plant.

Advantages and disadvantages of sexual and asexual reproduction

	Advantages	Disadvantages
Sexual reproduction	Produces variation in the offspring If the environment changes, the variation gives a selective advantage by natural selection Natural selection can happen more quickly by using selective breeding (useful in food production)	Need to find two parents Takes more time and energy to grow offspring Offspring may have unfavourable characteristics
Asexual reproduction	Only one parent needed More time- and energy-efficient Many identical offspring can be produced quickly	No variation in offspring Vulnerable to changes in the environment All offspring could be affected by a disease

Organisms that use sexual and asexual reproduction

Some organisms reproduce using both sexual and asexual methods. For example:

- **malarial** parasites reproduce sexually inside the mosquito, but asexually inside the human host
- many species of fungi reproduce sexually to produce variation, but also asexually by sending their **spores** away from them, usually carried by the wind or insects
- plants reproduce sexually using pollen and egg cells, but some plants can also reproduce asexually using runners (e.g. strawberry plants), bulbs (e.g. daffodils) and tubers (e.g. potatoes).

NAILIT!

You will only be expected to know these examples.

WORKIT!

Compare and contrast sexual and asexual reproduction. (4 marks)

Sexual reproduction needs two parents, and asexual reproduction needs one parent. (1)

Gametes in sexual reproduction are produced by meiosis, in asexual reproduction cells divide by mitosis. (1)

In sexual reproduction, the offspring are not genetically identical to the parents, whereas in asexual replication the offspring are genetically identical to the parent. (1)

In sexual reproduction the offspring have genetic variety, and in asexual reproduction the offspring are genetically identical to each other. (1)

CHECKIT!

1. Give two examples of asexual reproduction.
2. What is a zygote?
3. What are the advantages of asexual reproduction over sexual reproduction?

Meiosis

SNAPIT!

Sketch the diagram of meiosis and recall the number of chromosomes in the parent and daughter cells. Remember to take a photo!

Meiosis is a type of cell division that produces gametes. The gametes are sperm and eggs in animals, and pollen and egg cells in plants.

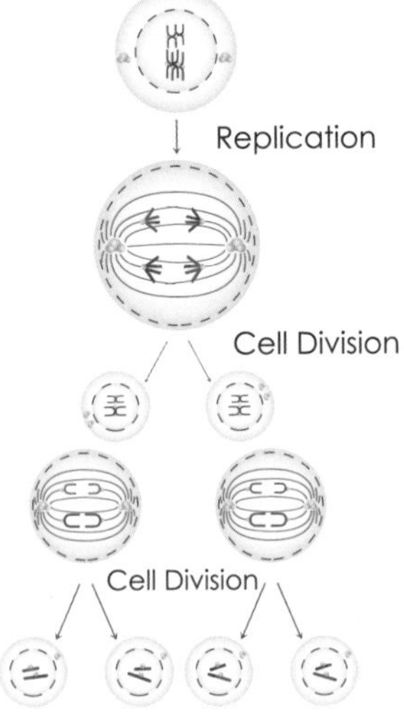

The process of meiosis

When a cell in the reproductive organs divides by meiosis, it goes through two rounds of cell division to produce four daughter cells. Each daughter cell has half of the number of chromosomes as the original body cell. The chromosomes are randomly assorted into the four daughter cells so that they are genetically different from each other. This is the cause of variation.

Fertilisation

Fertilisation happens when a male and a female gamete join together and the two nuclei fuse. When this happens, the new cell (the zygote) has a full set of chromosomes. The zygote divides by a type of cell division called mitosis (see page 18). The zygote divides many times to form an embryo. All of the cells in the embryo are genetically identical. As the embryo develops, the cells begin to differentiate into specialised cells.

NAILIT!

You will not be expected to know the stages of meiosis.

WORKIT!

Describe what happens during meiosis. (3 marks)

A cell in a reproductive organ goes through two rounds of division. (1)

The number of chromosomes in the daughter cells is halved. (1)

Chromosomes are randomly assorted into daughter cells to cause variation. (1)

CHECKIT!

1. An animal's body cells contain 12 chromosomes. How many chromosomes will the gametes contain?
2. What is mitosis?
3. Explain how zygotes contain the same number of chromosomes as the body cells of that organism.

DNA and the genome

DNA (deoxyribonucleic acid) is the genetic material of the cell.

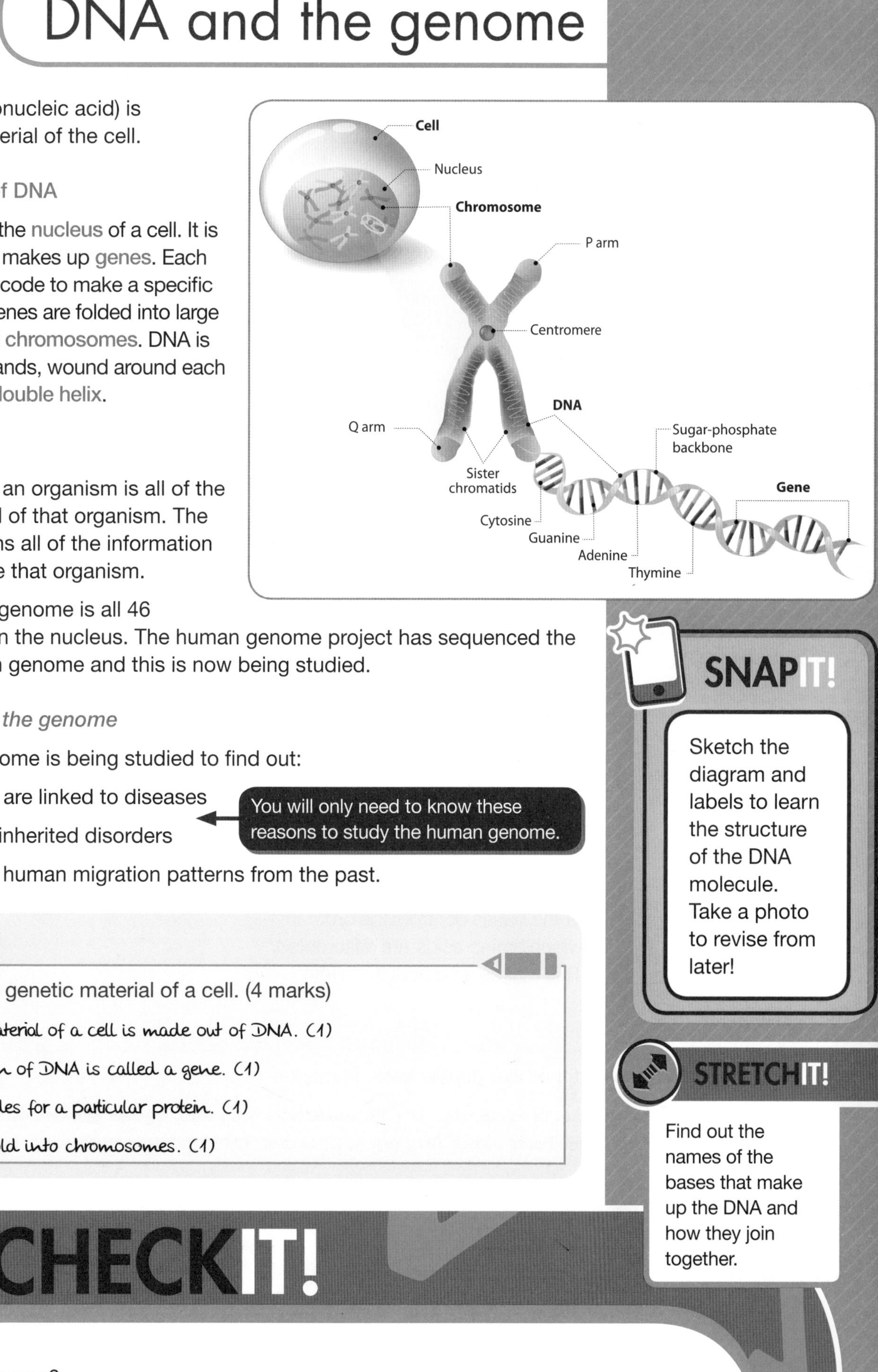

The structure of DNA

DNA is found in the nucleus of a cell. It is the material that makes up genes. Each gene contains a code to make a specific protein. Many genes are folded into large structures called chromosomes. DNA is made of two strands, wound around each other to form a double helix.

The genome

The genome of an organism is all of the genetic material of that organism. The genome contains all of the information needed to make that organism.

In humans, the genome is all 46 chromosomes in the nucleus. The human genome project has sequenced the whole of human genome and this is now being studied.

Understanding the genome

The human genome is being studied to find out:

- which genes are linked to diseases
- how to treat inherited disorders
- how to trace human migration patterns from the past.

You will only need to know these reasons to study the human genome.

SNAPIT!

Sketch the diagram and labels to learn the structure of the DNA molecule. Take a photo to revise from later!

WORKIT!

Describe the genetic material of a cell. (4 marks)

The genetic material of a cell is made out of DNA. (1)

A small section of DNA is called a gene. (1)

Each gene codes for a particular protein. (1)

Many genes fold into chromosomes. (1)

STRETCHIT!

Find out the names of the bases that make up the DNA and how they join together.

CHECKIT!

1. What is a genome?
2. Describe the shape of the DNA molecule.
3. Explain the importance of studying the human genome.

SNAPIT!

Sketch both diagrams to learn the structure of a nucleotide and the four bases. Take a photo of your sketches to revise from later.

DNA structure

DNA is a polymer made of many nucleotides joined together.

Nucleotides

A nucleotide consists of a sugar, a phosphate group and one of four different bases.

The nucleotides join together between the sugar on one nucleotide and the phosphate group on another nucleotide. This is called the sugar-phosphate backbone. DNA is made of many nucleotides joined together to make a polymer called a polynucleotide.

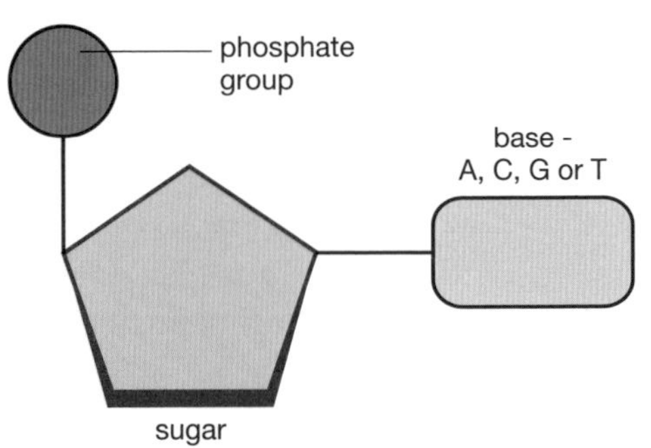

The four different bases in DNA are:

adenine (A); cytosine (C); guanine (G); thymine (T)

Adenine on one strand of DNA binds to thymine on the other strand of DNA with two hydrogen bonds. Cytosine on one strand of DNA binds to guanine on the other strand of DNA with three hydrogen bonds. This is called complementary base pairing.

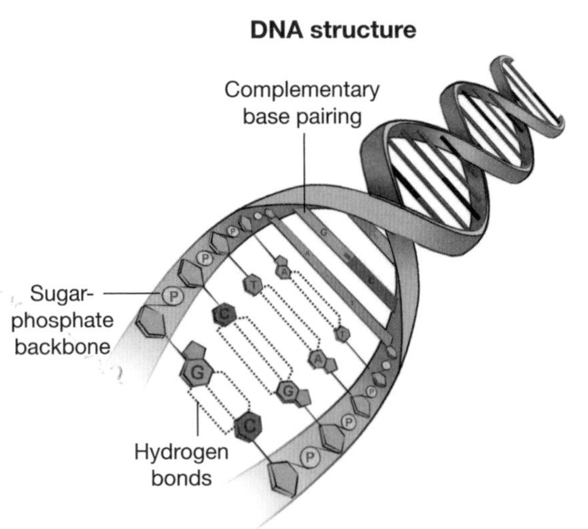

NAILIT!

You will only be expected to remember the letters of the bases, not the names.

STRETCHIT!

Find out how proteins are synthesised in the cell, using the DNA code.

The DNA code

A sequence of three bases codes for a particular amino acid. The order of the bases controls the order in which amino acids are assembled. This makes a particular protein.

WORKIT!

Describe the structure of the DNA double helix. (4 marks)

DNA is made of a long chain of nucleotides. (1) The nucleotides are joined together between the phosphate group and the sugar. (1) The bases of the DNA are A, C, G and T. (1) A binds to T and C binds to G. (1)

CHECKIT!

1. What is a nucleotide?
2. How do the bases join together?
3. Describe how the DNA code is used to make different proteins.

Protein synthesis

Proteins are synthesised on the ribosomes of the cell.

How proteins are synthesised

Protein synthesis starts in the nucleus, where each gene codes for a specific protein.

1 In the nucleus, the DNA code of the gene is transcribed into a messenger RNA (mRNA) code.
2 The mRNA leaves the nucleus and finds a ribosome.
3 The ribosome translates the code on the mRNA, three bases at a time.
4 Carrier molecules called transfer RNA (tRNA) bring the amino acids to the ribosome.
5 The three bases tell the ribosome which amino acid to add onto the protein next.
6 The completed protein breaks away from the ribosome and folds into its unique shape.

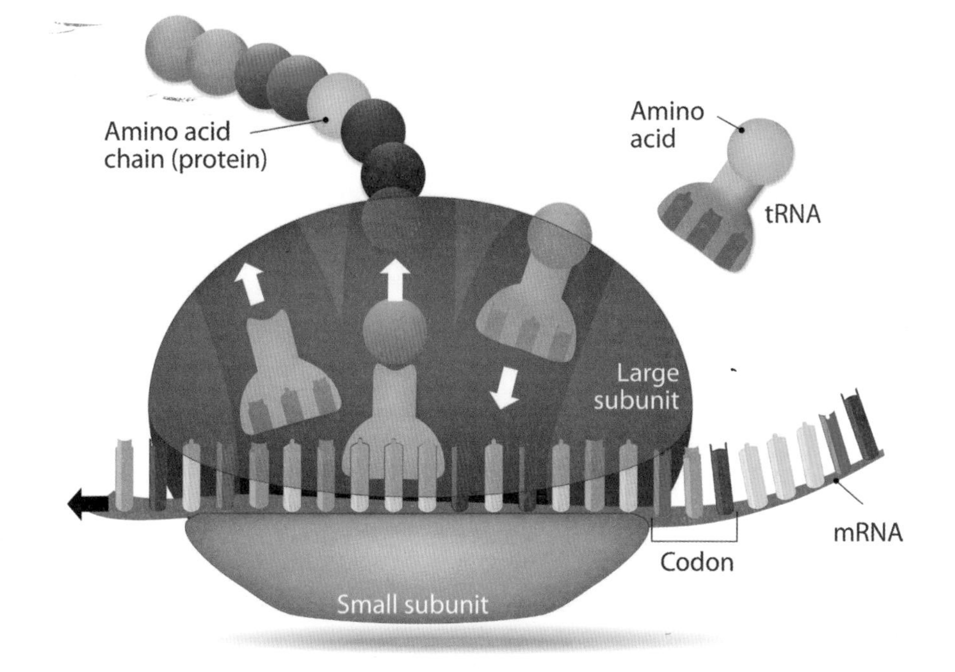

How proteins fold

When proteins are first synthesised, they are a long chain of amino acids. This chain folds into a 3D shape, depending on the sequence of amino acids. This means that each type of protein has its own unique shape. This is useful and enables proteins to work as enzymes, hormones or structural proteins.

Use modelling clay to model protein synthesis, using different colours for different amino acids.

You will not be expected to know the structure of mRNA or tRNA, or the detailed structure of amino acids or proteins.

H

Mutations

A mutation is a change in the order of bases in the DNA sequence. Mutations happen continuously in the cells, however the cells can repair most of them.

Mutations in coding DNA

The coding DNA codes for proteins. A change of even one base in the coding DNA can alter the three base code, which can alter the amino acid sequence of a protein. Most of the time, this does not alter the protein, or only alters it slightly, so the protein can still carry out its function.

Sometimes a mutation can affect the shape of a protein. This means that the protein will no longer be able to function. For example, enzymes have an active site, with a specific shape that is complementary to its substrate. If the shape of the active site is changed, then the substrate can no longer bind to the enzyme.

Find out about some common mutations in humans.

Mutations in non-coding DNA

The non-coding DNA controls the genes, switching them on and off. A change in the base sequence in non-coding DNA affects how the genes are expressed – whether or not the gene is switched on or off.

WORKIT!

Explain how a change in the DNA sequence can lead to a change in proteins. (4 marks)

A mutation in the DNA sequence can lead to a change in the base sequence. (1)

The altered base sequence in the coding DNA can lead to a different sequence of amino acids. (1)

This can change the way in which the protein functions. (1)

A change in the base sequence in non-coding DNA can change whether or not a gene is switched on or off, and therefore whether or not a protein is expressed. (1)

H1 Describe the role of the mRNA.

H2 Why is it important that different proteins have different amino acid sequences?

H3 Suggest what would happen to a protein if a base close to the beginning of the DNA sequence were deleted by a mutation.

Genetic inheritance

You inherit some genes from your mother and some from your father. Many characteristics are controlled by many genes, but where the characteristic is controlled by a single gene, the pattern of inheritance can be worked out using genetic diagrams.

Genetics keywords

Gamete – a sex cell, for example, sperm or an egg, containing half of the genetic material.

Chromosome – a large structure in the nucleus, made of DNA.

Gene – a short section of DNA that codes for a specific protein.

Allele – a version of a gene.

Dominant – an allele that is always expressed.

Recessive – an allele that is not expressed when a dominant allele is present.

Homozygous – when two copies of the same allele are present.

Heterozygous – when two different alleles are present.

Genotype – the alleles that are present in the genome.

Phenotype – the characteristics that are expressed by those alleles.

Single gene inheritance

Some characteristics are controlled by a single gene. The genotype is shown by two letters, a capital letter to represent the dominant allele, and a lower case letter to represent the recessive allele. The phenotype depends on the alleles of the gene that are present.

DO IT!

Put each keyword on a revision card with the definition on the back and test yourself.

Fur colour in mice

The fur colour gene in mice controls whether the fur colour is black or brown. There are two alleles, a dominant one, which is labelled B, and a recessive one that is labelled b.

- Mice with two dominant alleles, BB – black fur
- Mice with one dominant and one recessive allele, Bb – black fur
- Mice with two recessive alleles, bb – brown fur.

Red-green colour blindness in humans

The gene for red-green colour blindness is found on the X sex chromosome. If a female has two recessive alleles, or a male has one recessive allele, then that person will have red-green colour blindness.

- Male with the dominant allele, XY – normal vision.
- Male with the recessive allele, X^cY – red-green colour blindness.
- Female with one dominant and one recessive allele, XX^c – normal vision.
- Female with two recessive alleles, X^cX^c – red-green colour blindness.

Predicting inheritance

If the genotypes of the parents are known, it is possible to work out the possible genotypes and phenotypes of the offspring. It is also possible to calculate the probability of each genotype and phenotype.

For example, if two black mice that have the genotype Bb breed together, the possible genotypes of the offspring are BB, Bb or bb. The probabilities of each genotype are 1:2:1.

Therefore, out of four potential offspring, three would have black fur, and one would have brown fur, in a ratio of 3:1.

STRETCH IT!

Practise working out genotypes and phenotypes and the probabilities of each for the offspring, with parents of different genotypes.

WORK IT!

Describe and explain how the phenotype of a characteristic is controlled. (4 marks)

Each gene has two alleles that are present. (1)

Dominant alleles are always expressed if at least one dominant allele is present. (1)

Recessive alleles are only expressed if both of the alleles are recessive. (1)

The expressed alleles give the phenotype. (1)

NAIL IT!

For Q2a work out the genotypes of the parents first.
The woman will have the genotype, X^cX and the man will have the genotype, XY.

CHECK IT!

NAIL IT!

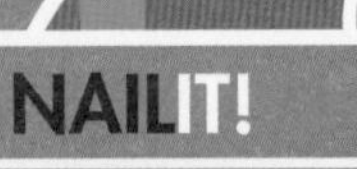

Remember, probability is represented as a number between 0 and 1, **not** a percentage (%).

1 What does 'homozygous dominant' mean?

2 a A woman who is heterozygous for red-green colour blindness has a child with a man with normal vision. What are the possible genotypes for the child?

b What is the probability that they will have a boy with red-green colour blindness?

Punnett squares

A Punnett square is a type of genetic diagram. You can use these to predict the genotypes of offspring.

How to use a Punnett square

Punnett squares can be used to work out the genotypes of offspring, if you know the genotypes of the parents. A small grid, 2 x 2 is drawn with the genotype of one parent on top of the grid, and the genotype of the other parent on the side of the grid.

Working out ratios

A ratio shows how many of the offspring are expected to have each phenotype. For example, in the Punnett square to the right, the ratio is 3:1. Three offspring would be expected to have the dominant characteristic and one offspring would be expected to have the recessive characteristic.

The letter at the top of each column is written in the boxes below. The letter at the side is written in each row. This gives four possible genotypes for the offspring:

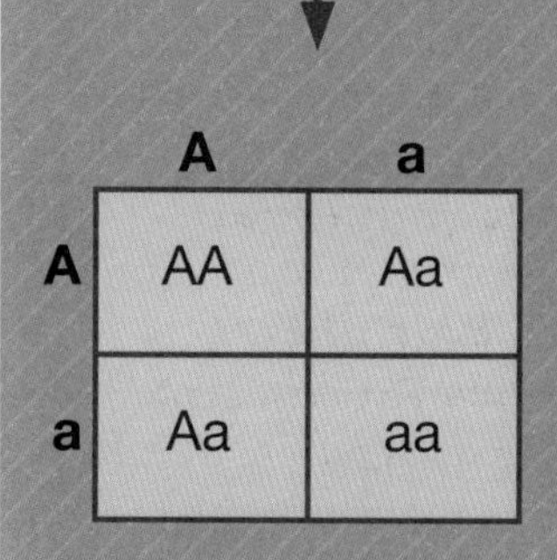

	A	a
A	AA	Aa
a	Aa	aa

WORKIT!

In some flowers, petal colour is controlled by a single gene. The dominant allele, R, gives the petals a red colour. The recessive gene, r, gives petals a white colour.

a What are the possible genotypes of the offspring if one parent has the genotype Rr, and the other parent has the genotype rr? (2 marks)

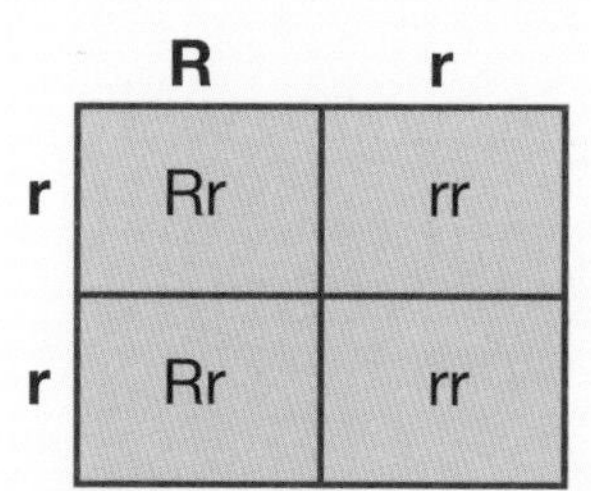

	R	r
r	Rr	rr
r	Rr	rr

The possible genotypes are Rr and rr. (1)

b What percentage of the offspring will have white petals? (1 mark)

50% (1)

c What is the ratio of phenotypes? (1 mark)

1:1 (1)

NAILIT!

All students need to be able to complete a Punnet square diagram. Only Higher Tier students need to be able to create their own Punnet square diagrams to make predictions about genetic crosses.

DOIT!

Practise using Punnett squares to work out the possible genotypes with parents of different genotypes. Try using RR and Rr, then Rr with rr. Are there any other combinations?

Family trees

It is possible to work out the genotypes of parents and offspring by looking at family trees.

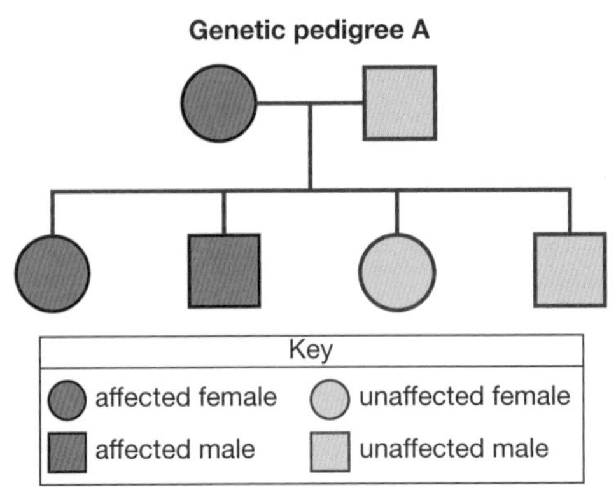

This example shows a disease caused by a single dominant allele. The mother and two of her children have one dominant allele and therefore have the disease. The father and the other two children have two recessive alleles and do not have the disease.

STRETCHIT!

Look at the pedigree for the British royal family and look at the inheritance of the gene for haemophilia.

WORKIT!

This family tree shows the inheritance of cystic fibrosis. People with two recessive alleles, cc, will have cystic fibrosis. What are the genotypes of the parent of the man with cystic fibrosis? Explain how you know this. (3 marks)

Both parents must have the genotype Cc. (1)

The parents need to have at least one recessive allele to pass on to their son. (1)

If they had two recessive alleles, they would have cystic fibrosis. (1)

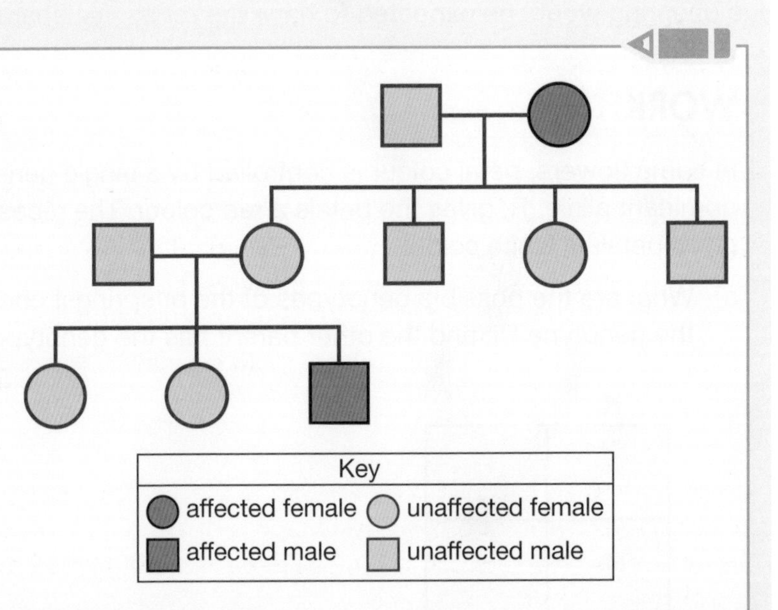

CHECKIT!

1 What is a dominant allele?

2 If the results of a Punnett square show that one offspring is homozygous dominant, two offspring are heterozygous, and one offspring is homozygous recessive. What is the ratio of their phenotypes?

H 3 In the above worked example, the woman with cystic fibrosis has four children. The father does not have a recessive allele for cystic fibrosis. What are the possible genotypes of the children?

NAILIT!

Use a Punnett Square to work out the answer to Q3.

Inherited disorders

Some disorders are caused by a single gene and can be inherited.

Disorders caused by a single gene

Inherited disorders can be caused by a dominant or a recessive allele. For example:

- polydactyl – having an extra finger or toe. Caused by a dominant allele
- cystic fibrosis – a disorder that affects the cell membranes of the lungs and pancreas. Caused by a recessive allele.

People who know that they carry an allele for a disorder may wish to go through genetic counselling. This is when they discuss the likelihood of passing on their allele. Some people wishing to have children may go through embryo screening.

Embryo screening

Embryos can be screened to see if they contain the allele for the disorder. Only embryos without the allele will be implanted into the mother's uterus.

Advantages and disadvantages

Advantages	Disadvantages
Can select embryos without the allele for a disorder	Embryos with the undesired allele are discarded Procedure is expensive Could be used for non-disease alleles

Sex determination

Human body cells have 23 pairs of chromosomes. 22 pairs of chromosomes are autosomal chromosomes and two are sex chromosomes (allosomal). Human females have two X chromosomes. Human males have an X and a Y chromosome.

The possible sex chromosomes of the offspring can be worked out using a Punnett square (see punnett square to the right).

The probability of the offspring having male or female sex chromosomes is 0.5.

WORKIT!

Describe and explain the economic, social and ethical issues concerning embryo screening. (4 marks)

Embryo screening allows parents to make sure that their child is not born with an allele for a disorder. (1) However, it is expensive so some people would not be able to afford it. (1) Some parents might want to discard embryos with other alleles/make designer babies. (1) Embryos that are not chosen are discarded, but could have become babies if they had been implanted into a uterus. (1)

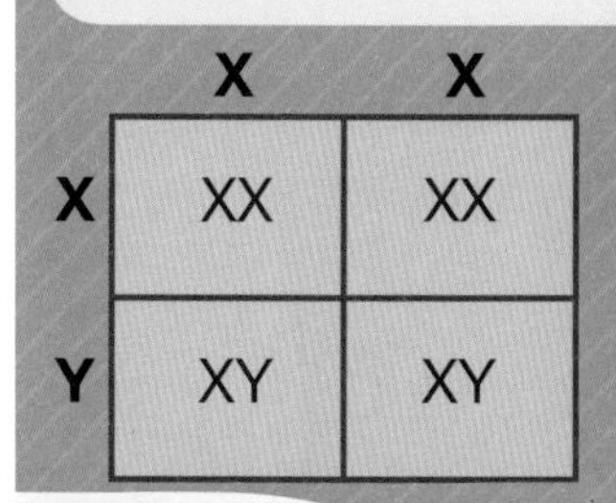

1 Name a disorder caused by a dominant allele.

2 Which sex chromosomes are present in human males?

3 A couple who are both heterozygous for cystic fibrosis want to have a child together. What is the chance they will have a child with cystic fibrosis?

DOIT!

Write out the advantages and disadvantages of embryo screening onto revision cards and learn them.

Variation

Organisms are different from one another. This is called variation.

Causes of variation

There is extensive genetic variation within a population of a species.

The two causes of variation are:

1. Genetic – the genes you inherit from your parents control many aspects of your phenotype. For example, your hair and eye colour.
2. Environmental – the world around you can have an influence of some aspects of your phenotype. For example, if you live in a hot climate, you may develop more pigmentation in your skin.

Many phenotypes are influenced by both genetic and environmental factors, for example, height.

Mutations

Mutations are a change in the base sequence of the DNA (see page 104) and happen continuously. Most mutations are repaired by the cell, but some remain in the genome. All variants arise from mutations. Most of the time, the new variant, for example, a change in a protein, will have no effect on the phenotype of the organism. Some variants will influence the phenotype. Rarely, a variant will lead to a new phenotype.

If the new phenotype is advantageous to the organism in its environment, the mutation will spread throughout the population by natural selection.

If the new phenotype is not advantageous to the organism, then that organism is less likely to have offspring and the mutation will not spread through the population.

DO IT!

Make a table of phenotypes such as hair colour, eye colour, hair length and height, and decide whether they are caused by genetics only, environment only, or both.

WORK IT!

Explain why variation in organisms of the same species is important. (3 marks)

If all organisms are genetically identical, they will all be susceptible to the same diseases. (1) Variation allows some individuals to survive better than others if the environment changes, (1) through the process of natural selection. (1)

CHECK IT!

1 Give an example of a phenotype that is influenced by genes and the environment.

2 What is a mutation?

3 Describe what happens to a mutation that gives an individual an advantage in its environment.

Evolution

Evolution is the gradual change in inherited characteristics of organisms over time through natural selection.

Natural selection

Within any population there is variation. Some individuals have characteristics that are better suited to the environment and are more likely to survive. Therefore, those individuals are more likely to have offspring and pass on those characteristics. This is natural selection.

Evolution

Life started on the Earth as very simple life forms more than three billion years ago. Since then, those simple life forms have gradually evolved into more complex life forms, through the process of natural selection.

Speciation

A species is a group of similar individuals that can breed together to produce fertile offspring. If the phenotypes of two populations of a species become so different that they can no longer interbreed, then they are no longer the same species, but two different species. This process of forming new species is called speciation.

Variation | Natural selection | New generations

Individuals show variation. Some offer an advantage. | Best suited are favoured and selected. | Variation increases in frequency.

SNAPIT!

Make your own version of the diagram and learn the process of natural selection. Take a photo to revise from later.

WORKIT!

A population of rabbits that lived in the Arctic had individuals with brown or white fur. Arctic wolves were eating the rabbits. The same population today has mostly individuals with white fur. Explain the mechanism that brought about these changes. (4 marks)

The rabbits with white fur were camouflaged in the snow. (1) They were less likely to be eaten by the arctic wolves. (1) They passed on the white fur characteristic to their offspring. (1) This is natural selection. (1)

STRETCHIT!

Find out about Darwin's finches and how they became separate species.

CHECKIT!

1 What is a species?

2 Describe what would happen to individuals in a population who do not have advantageous characteristics.

3 Suggest how speciation would occur in two populations of a species that have been isolated from one another.

NAILIT!

For Q3 remember to include natural selection in your answer.

Selective breeding

NAILIT!

Selective breeding is also called artificial selection because plants or animals that do not have the desired characteristics are not allowed to breed.

STRETCHIT!

Research selective breeding in wheat and see how modern wheat looks compared with the ancestral type.

Food crops and domesticated animals have been selectively bred for thousands of years.

The process of selective breeding

Selective breeding is the breeding together of organisms with desired characteristics to produce offspring with the same desired characteristics. Selective breeding can lead to inbreeding where some breeds are particularly prone to disease or inherited defects. This can be seen in some breeds of dog.

Food plants

Two plants with the desired characteristics are selected. The pollen from one plant is transferred to the other plant for sexual reproduction. The resulting seeds are planted and grown into new plants. The best of these plants are bred together to make new plants. This continues for many generations until all of the plants produce offspring with the desired characteristics. For example, cauliflower plants that produce large cauliflower heads.

WORKIT!

A gardener wants to produce tomato plants with larger tomatoes. Describe how the gardener could achieve this. (3 marks)

Choose two tomato plants that produce the largest tomatoes. (1) Breed these two plants together. (1) Choose the offspring that produce the largest tomatoes and breed them together. (1)

Domesticated animals

A good example of selective breeding in domesticated animals is milk production in cows. A cow that produces a high yield of milk is bred with a bull that produces daughters with a high yield of milk. The daughters that produce the highest yield of milk are bred with other bulls that have produced high yield daughters. This process continues for many generations until all female offspring produce high yields of milk.

Desirable characteristics include:

- disease resistance in food crops e.g. blight-free potato plants
- large/unusual flowers
- animals that produce more meat/milk
- domestic dogs with a gentle nature.

The benefits and risks of selective breeding

Benefits	Risks
All offspring have desired characteristics Can eliminate diseases Increases the productivity/yield	Takes a long time to produce organisms with desired characteristics Loss of variety in species Animals may suffer discomfort, for example, cows with heavy udders Risk of inbreeding if two animals are too closely related

DOIT!

Write the benefits and risks of selective breeding onto revision cards.

CHECKIT!

1 Give an example of selective breeding.

2 Name two benefits of the selective breeding of plants.

3 Explain why selective breeding is also known as artificial selection.

Genetic engineering

What is genetic engineering?

Genetic engineering is the addition of a gene from another organism into an organism's genome.

Plants and animals have been selectively bred over many generations in order to produce organisms with desired characteristics. Genetic engineering aims to achieve the same outcome over one generation by adding a gene to that organism from another organism, usually from another species. The desired gene is cut out from the genome of the original organism and inserted into the organism to be modified using a number of different techniques. Organisms that have been modified in this way are called genetically modified organisms, or GMO.

Examples of genetic engineering:

- disease-resistant plants
- plants that produce bigger fruits
- bacteria that produce human insulin (for people with type 1 diabetes)
- gene therapy to overcome some inherited disorders, e.g. severe combined immunodeficiency (SCID).

Benefits and risks of genetic engineering

Benefits	Risks
Can produce organisms with the desired characteristics more quickly GMOs increase yield/productivity Food plants can be more nutritious, e.g. golden rice Crops can be grown in harsher environments Crops can be made resistant to pesticides Medicines can be made more cheaply, e.g. human insulin	Eating GMOs could potentially cause harm to people Can spread inserted gene to other species GM crops could cause harm to insect species Difficult to get the inserted gene into the right place in the genome

GM crops

Crops that have been genetically modified are called GM crops. Some examples of GM crops are:

- drought-resistant crops that can be grown in very dry conditions
- pest-resistant crops that contain genes that kill the pest
- herbicide-resistant crops
- crops that have had vitamins added, e.g. beta carotene (the precursor to vitamin A) can be added to golden rice
- crops that have had genes added that make the fruit last longer.

GM crops have a greater yield and therefore can feed more people. However, some people are concerned that eating GM crops will have unforeseen health consequences. Others also fear that genes such as pesticide-resistance genes will pass to weeds, causing them to also be resistant to pesticides. Pest-resistant plants may also cause harm to other insects, besides the pest, that eat the plant.

STRETCH IT!

Find out about the research into gene therapy for cystic fibrosis.

SNAP IT!

H

Sketch a copy of the genetic engineering diagram and add your notes to describe each stage. Take a photo to revise from later.

Gene therapy

Gene therapy is the insertion of a gene into the genome of a person who has a disease or disorder, in order to remove the disease. For example, severe combined immunodeficiency (SCID) is a disorder that causes the immune system to no longer function. It is caused by a faulty **allele**. Gene therapy has successfully replaced the faulty allele in children with SCID with a normal allele, eliminating the disease. Trials to replace the faulty alleles of other diseases, such as cystic fibrosis, have proved difficult because it is difficult to get the gene into the correct place in the genome.

The process of genetic engineering

H

A common method of inserting a gene into a genome is to use **vectors**. These are usually a plasmid or a virus. Plasmids are small circles of DNA, originally found in bacteria.

1. The desired gene is cut out of the original genome using enzymes called **restriction enzymes**.
2. The same enzymes are used to cut open a vector.
3. The gene is inserted into the vector.
4. The vector is inserted into the cells of animals, plants or microorganisms at an early stage in their development. The organisms will develop with the desired characteristic.

This diagram shows the desired gene being inserted in bacteria.

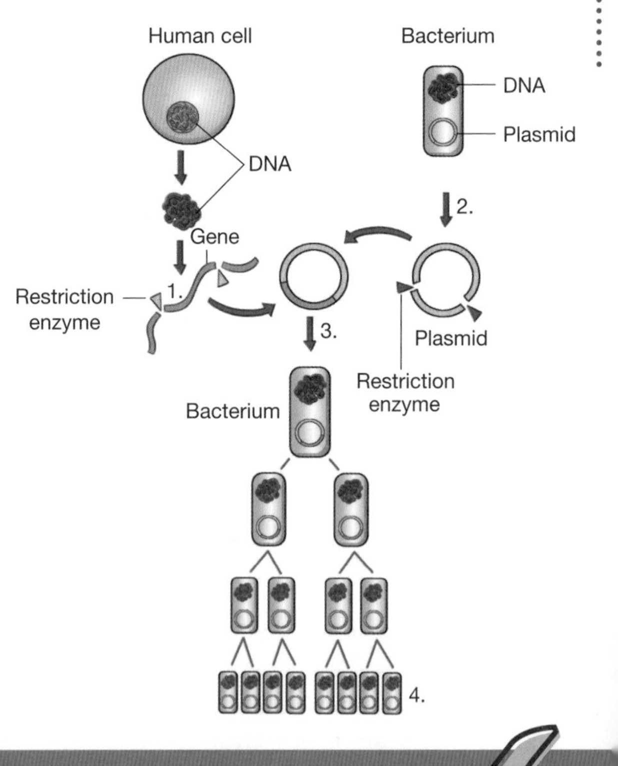

WORK IT!

Compare and contrast the benefits and risks of genetic engineering in plants. (4 marks)

Food plants can be made more nutritious, but it can be difficult to get the gene into the correct place in the genome. (1)

Can make plants resistant to herbicides, but there is the possibility of passing on this resistance gene to weeds. (1)

Can make plants resistant to pests, but could kill other insects not just the pests. (1)

GM crops produce a greater yield, but could potentially cause harm in people. (1)

CHECK IT!

1. Give an example of genetic engineering.
2. Describe simply how an organism is genetically modified.
3. Explain how gene therapy could be used to treat people with a disorder caused by a faulty allele.

Cloning

Cloning often occurs in nature but can be used artificially to produce many identical organisms.

The processes of cloning

There are several different methods of cloning; some are easily achieved with a few simple tools, others are more complicated and are carried out in a laboratory.

Cuttings

Gardeners have been using cuttings for many generations to produce new plants. The gardener uses a sharp knife to take a section of a plant. The plant is dipped in rooting hormones and placed into soil where it will grow roots. Many identical plants can be produced in this way.

Tissue culture

Tissue culture is carried out in a sterile environment, to prevent fungi or bacteria from growing on the agar plate. A small group of cells from a plant is placed onto a sterile agar plate containing nutrients and plant hormones called auxins. The cells grow into small plantlets, and are then placed into soil where they grow into plants. This technique works well with plants because the cells can differentiate into any specialised cell type.

SNAP IT!

Make a step-by-step list to describe the process of adult cell cloning and take a photo to help you remember.

Embryo transplants

This type of cloning is used in animals. The egg is fertilised and allowed to grow into an embryo. The embryo is split before the cells have begun to differentiate, into several smaller embryos. These embryos are then implanted into uteruses and grow into genetically identical animals. This technique allows more offspring with desired characteristics to be made at the same time.

Adult cell cloning

Adult animals can be cloned using the adult cell cloning technique. The first animal to be made in this way was Dolly the sheep in 1996.

1. An adult body cell is removed from one animal, and an egg from another animal.
2. The nucleus is removed from the egg.
3. The nucleus from the body cell is inserted into the egg.
4. An electric shock stimulates the egg to divide to form an embryo.
5. These embryo cells contain the same genetic information as the adult body cell.
6. When the embryo has developed into a ball of cells, it is inserted into the uterus of an adult female, where it grows into an animal that is a clone of the animal that donated a body cell.

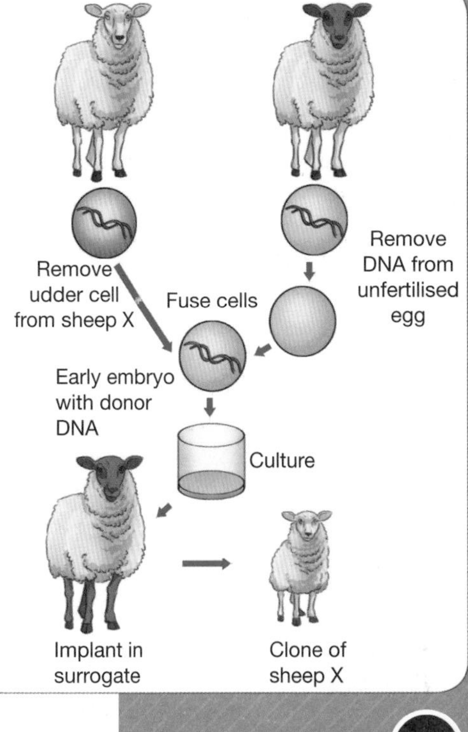

Benefits and risks of cloning

Benefits

- It is possible to produce many organisms with the desired characteristics.
- It is possible to make large numbers of more rare species.

Risks

- All clones will be susceptible to the same diseases.
- The clones will have no variation, and so will not be able to respond to changes in the environment.
- The lack of variation makes it more difficult to make new varieties in the future.

STRETCHIT!

Find out about Dolly the sheep and some of the difficulties in making her.

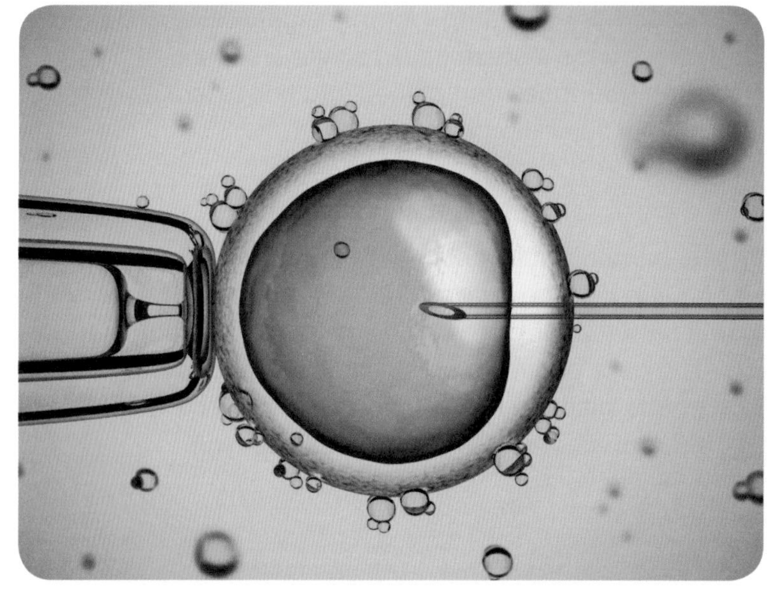

WORKIT!

Compare and contrast the methods of embryo transplants and adult cell cloning. (3 marks)

Adult cell cloning produces one clone, but embryo transplants produce many clones. (1)

Both techniques use surrogate mothers. (1)

The embryos in embryo transplants are produced by sexual reproduction, whereas the DNA in the egg in adult cell cloning is taken from a body cell. (1)

CHECKIT!

1. Name a plant cloning technique that uses hormones.
2. Why does tissue culture have to be carried out in a sterile environment?
3. Explain why a population of cloned organisms is susceptible to a change in the environment.

Theory of evolution

Charles Darwin published his theory of evolution by natural selection in 1859.

Charles Darwin

Charles Darwin went on a five-year voyage on a ship called HMS *Beagle*. On his travels he observed and collected many animal specimens and fossils. He also observed the geology of the lands he visited. Once back home, he discussed his findings with other scientists and conducted experiments on selective breeding using pigeons. It was after this that he proposed his theory of evolution by natural selection, in his book, *On the Origin of Species*.

The theory of evolution by natural selection

During his travels, Darwin made the following observations:

- Individual organisms within a particular species show a wide range of variation for a characteristic.
- Individuals with characteristics most suited to the environment are more likely to survive to breed successfully.
- The characteristics that have enabled these individuals to survive are then passed on to the next generation.

Darwin noted that many more offspring are born than survive, and for many species life is a struggle to find food and escape from predators. These struggles are called selection pressures. It is for this reason that natural selection is often called 'survival of the fittest'.

On the Origin of Species

At the time when Darwin's book was published, the idea of evolution was very controversial. The general view at the time was that all species had been created by God and were the same today as they had always been. There was also not yet enough evidence to support the theory of evolution to convince many scientists, and genetics was not yet understood. The mechanism of the inheritance of genes was not known for 50 years after the theory was published. For these reasons, the theory of evolution was only gradually accepted.

There were other theories of evolution at the same time as Darwin's. For example, Jean-Baptiste Lamarck proposed the theory that changes that happened to an organism in its lifetime, could be passed onto its offspring. This kind of inheritance, however, does not occur. For example, a body builder does not pass on their muscles to their offspring.

STRETCHIT!

Find out about some of the finch specimens that Darwin brought back from the Galapagos Islands.

Use an online natural selection simulation to see how natural selection works.

WORKIT!

What types of evidence did Darwin use before proposing his theory of evolution by natural selection? (4 marks)

Collected specimens and fossils. (1)

Examined the geology of the land. (1)

Carried out experiments on selective breeding. (1)

Discussed his work with other scientists. (1)

1 Besides natural selection, name one other theory of evolution.

2 What is a selection pressure?

3 Explain why Darwin's theory was controversial.

Speciation

STRETCH IT!

Research the work of Alfred Russel Wallace on speciation.

SNAP IT!

Draw your own diagram to explain the two types of speciation and snap a photo to revise from later.

Speciation is the formation of a new species.

Darwin and Wallace

Alfred Russel Wallace came up with the theory of evolution by natural selection independently of Charles Darwin. Wallace and Darwin published joint papers on the topic in 1858, before Darwin published his book, *On the Origin of Species*, in 1859. Wallace worked around the world gathering evidence for evolution and is known for his work on warning colouration in animals and speciation.

The process of speciation

In order for a new species to emerge, two populations of the same species must be separated. This can be:

- geographical separation – populations that live in different geographical locations, e.g. different islands. This is called allopatric speciation.
- reproductive isolation – populations cannot breed together due to different mating seasons or courting behaviours. This usually happens in the same geographical area and is called sympatric speciation.

Allopatric speciation

Sympatric speciation

When two populations are separated, the following steps occur:

1. The two populations will have different selection pressures
2. The phenotypes of the two populations will change due to natural selection
3. The two populations will no longer be able to interbreed and produce fertile offspring
4. The two populations are now considered to be separate species.

WORK IT!

Two populations of a finch species live on different islands. One population feeds on insects while the other feeds on seeds. Explain how these two populations could become two separate species. (3 marks)

The two populations would not interbreed due to geographical separation. (1) Each population has different food sources. (1) The phenotypes of each population would change due to natural selection. (1) Eventually the two populations would no longer be able to interbreed with each other to produce fertile offspring. (1)

CHECK IT!

1. Give two types of separation between populations.
2. Describe Alfred Russel Wallace's contribution to the theory of evolution by natural selection.

The understanding of genetics

The first person to show the pattern of inheritance was Gregor Mendel in the mid-19th century.

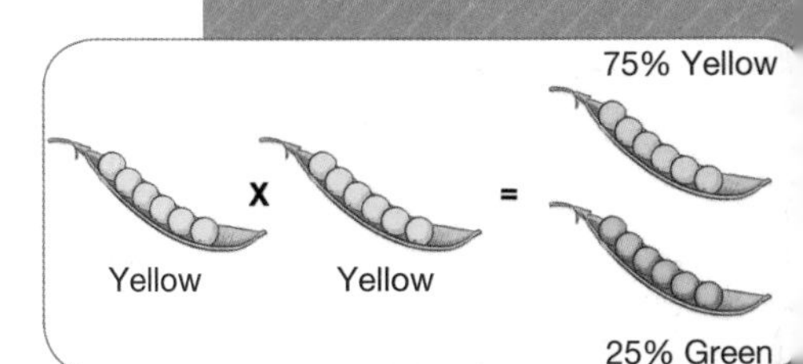

Mendel's peas

Gregor Mendel was an Austrian monk who carried out experiments on pea plants in his garden in order to observe the pattern of inheritance. He noticed that certain characteristics were inherited more frequently than others.

Mendel observed that when breeding together two parent plants showing the more frequent characteristic, he often saw the more frequent and less frequent characteristics in the ratio 3:1. He stated that each characteristic was determined by a 'unit' that was passed on to offspring unchanged. We now know that those units are genes. For this reason Mendel is often called the father of genetics.

We now know that the more frequent characteristic, the yellow pea, is caused by a dominant allele and the less frequent characteristic, the green pea, is caused by a recessive allele.

The importance of Mendel's discovery

When Mendel was carrying out his experiments, genetics was very poorly understood. Chromosomes were discovered within cells in the late 19th century, and the behaviour of chromosomes during cell division observed using a microscope.

In the early 20th century, it was observed that chromosomes behaved in the same way as Mendel's 'units'. It was hypothesised that the 'units' were part of the chromosomes. In the 1950s, the structure of DNA was worked out. This lead to the discovery of genes and the discovery that 'units' are actually genes. By this time, Mendel had died, but his observations of the patterns of inheritance were vital to the development of gene theory.

STRETCHIT!

Find out about the seven different characteristics that Mendel observed in pea plant experiments.

DOIT!

Use a Punnett square to show the inheritance of Mendel's pea colour.

WORKIT!

Mendel bred together two pea plants that produced yellow peas. Three of the offspring also produced yellow peas, and one offspring produced green peas. Explain why. (4 marks)

The yellow pea colour is caused by a dominant allele and the green pea colour is caused by a recessive allele. (1) Both of the parents have one dominant allele and one recessive allele. (1) Three offspring have at least one copy of the dominant allele and so have yellow peas. (1) One offspring has two copies of the recessive allele and so has green peas. (1)

CHECKIT!

1 What did Mendel discover?

2 What causes the more frequent characteristic observed by Mendel?

3 Explain why the mechanism of inheritance was not understood in the mid-19th century.

Evidence for evolution

The theory of evolution has a lot of evidence to support it, such as the fossil record, antibiotic resistance in bacteria, and knowledge of genetic inheritance. The theory of evolution is now widely accepted.

The fossil record

Evolution is the gradual change of simple life forms into more complex ones. This can be seen in the fossil record. Fossils are the remains of organisms from millions of years ago and can be found in rocks. The oldest, simplest fossils are found in the oldest rocks. More complex fossils are found in the newer rocks.

How fossils are made

Fossils are formed if:

- parts of the organism's body, e.g. shell or bones, have not decayed because the organism died in a place where one or more of the conditions needed for decay are absent
- parts of the organism, e.g. shell or bark, are replaced by minerals as they decay
- traces of the organism, e.g. footprints or leaf prints, are preserved.

The fossil record is incomplete. This is because the soft part of an organism's body often decays, which means that there are few traces of many early life forms. Geological activity, such as earthquakes and volcanic eruptions have destroyed many fossils as well.

STRETCH IT!

Find out about whole preserved organisms such as woolly mammoths found in the ice.

Resistant bacteria

Bacteria developing antibiotic resistance shows the mechanism of natural selection in a short time frame. Bacteria divide rapidly and can mutate rapidly as well. Any mutations can be quickly passed on to offspring, and to nearby bacteria by a process called conjugation.

There is variation in any population of bacteria. When a population of bacteria is exposed to an antibiotic, most of the population will die. If there is a bacterium with a mutation that gives it resistance to the antibiotic then that bacterium will survive, and pass its mutation to its offspring. As the mutation is spread through the population, the population will gradually become resistant to that antibiotic. We call this is a resistant strain. One strain of bacteria called MRSA (methicillin-resistant *Staphylococcus aureus*) is resistant to many types of antibiotic.

To reduce the rate of antibiotic-resistant strains of bacteria developing we should:

- make sure that antibiotics are not prescribed inappropriately, e.g. to treat a virus
- make sure that a course of antibiotics is finished, even if the person is feeling better. This kills all of the bacteria so none survive and form resistant strains
- restrict the use of antibiotics in agriculture.

Scientists can develop new antibiotics that there is no resistance to. However, this takes a long time and costs a lot of money. It is difficult to make new antibiotics at the speed at which the bacteria are becoming resistant to them.

Evolutionary trees

Organisms can be organised into evolutionary trees based on their DNA sequences, physical characteristics and behaviour. The more similar the species are, the more closely related the species are.

Where branches occur, speciation has taken place (see page 118). Organisms that are close together on a branch are more closely related. In this evolutionary tree, species A and B are very closely related, and E and G are less closely related.

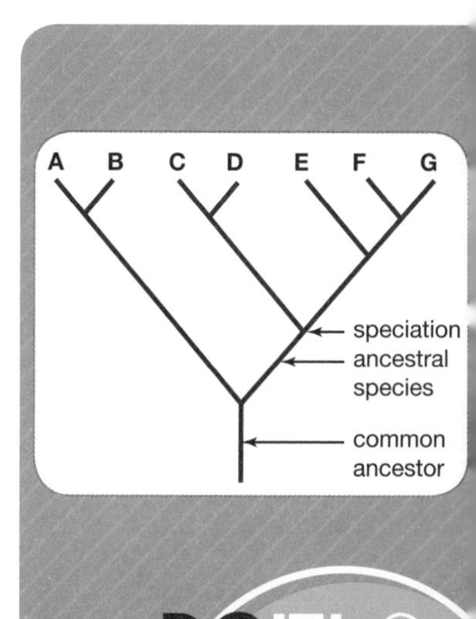

Extinction

When there are no longer any members of a species left on Earth, that species is extinct. Some examples of extinct species are the dinosaurs and woolly mammoths. We know that these species once existed because we can see their bodies in the fossil record. There are many reasons why a species may become extinct: loss of habitat, loss of food sources, hunting of the species.

DOIT!

Look at evolutionary trees in your textbook and online and work out how the species are related.

WORKIT!

This evolutionary tree shows the evolution of five species.

shark salamander lizard armadillo human

a Which two species are the most closely related? (1 mark)

Armadillo and human (1)

b Which two species are the least closely related? (1 mark)

Shark and human (1)

c Describe how the evolutionary tree is worked out. (2 marks)

The DNA sequences, physical characteristics, and behaviour of each species is compared. (1)
The more similar two species are, the more closely related they are. (1)

1. What is a fossil?
2. Give two reasons why a species might become extinct.
3. Explain how antibiotic resistance in bacteria shows the mechanism of natural selection.

Classification

DO IT!

Use a mnemonic to learn the order of the groups, e.g. King Phillip Came Over From Great Spain.

All organisms are classified into groups based on their physical characteristics.

The Linnaean system

In the 18th century, Carl Linnaeus proposed the organisation of all species into groups:

Group		Example
Kingdom	→	Animal
Phylum	→	Vertebrate
Class	→	Mammal
Order	→	Carnivore
Family	→	Felidae
Genus	→	Panthera
Species	→	Leo

Linnaeus put the species into groups depending on their structure and characteristics.

STRETCH IT!

Have a look at similar species and see how they are classified using the Linnaean system.

The binomial system

Linnaeus also proposed a system of giving all organisms two names to describe their genus and species. For example, the big cats all belong to the genus, *Panthera*, and so all have the same first name. However, different species of big cat have a different second name. So lions are called *Panthera leo*, and tigers are called *Panthera tigris*. This two-name system is called the binomial system, and is written in italics.

New models of classification

In the 20th century, technological advances have meant that scientists have a better understanding of:

- the internal structures of the cell using microscopes
- biochemical processes, such as the amino acid sequences of key proteins
- genetics, including the sequencing of genomes.

This means that people can compare cell structures, amino acid sequences in proteins, and DNA sequences in different species and see how closely related they are.

Three-domain system

In the late 20th century, Carl Woese proposed the three-domain system of classification. In this system all organisms are organised into one of three domains:

Archaea – primitive bacteria that live in extreme environments such as volcanoes

Bacteria – true bacteria

Eukaryota – all animals, plant, fungi and protists.

Using biochemical analysis, Woese found that archaea bacteria were evolutionarily more closely related to eukaryotes than bacteria, and should therefore have their own domain.

WORKIT!

Describe how new technologies have changed the way that we classify organisms. (3 marks)

We can observe the internal structures of cells, and work out the amino acid sequence in proteins, and DNA sequences. (1)

We can use this information to see how closely related species are. (1)

New three-domain system proposed. (1)

1 Fill in the missing groups.

kingdom ________ class ________ ________ genus species.

2 What is the binomial system?

3 Explain why archaea and bacteria are in two different domains.

REVIEW IT! Inheritance, variation and evolution

1 a Define asexual reproduction.
  b Compare daughter cells made by mitosis and meiosis.
  c Describe fertilisation.

2 a What is a gene?
  b What is a chromosome?
  c Describe the structure of a DNA molecule.
  d A DNA molecule was made up of 26% A (adenine) bases.
    i What percentage of the DNA molecule was made up of T (thymine) bases?
    ii What percentage of the DNA molecule was made up of C (cytosine) bases?
  e A section of DNA on one strand of the DNA molecule has the bases:
    ATGCCGTTAATC
    What are the complementary bases on the opposite strand?

3 a Name the two causes of variation.
  b Explain how variation leads to natural selection.

4 a What is meant by binomial classification?
  b Complete the table to show the classification of the Wolf, *Canis lupis*

Kingdom	Phylum		Order		Genus	Species
	Vertebrate	Mammal	Carnivore	Canidae		lupis

5 a The colour of fur in mice is controlled by a single gene. The dominant allele, B produce black fur. The recessive allele, b produces brown fur.
    i Use a Punnett square to work out the genotypes of the offspring of two black mice with the genotype, Bb.
    ii What is the ratio of the genotypes?
  b Red-green colour-blindness is a recessive disorder inherited on the X chromosome. If a colour-blind woman and a man who is not colour-blind had a child, what is the probability that their child will be colour-blind?

H 6 a What is genetic engineering?
  b Describe how human genes can be inserted into bacteria.
  c Evaluate the advantages of genetic engineering of human genes into other organisms.

Communities

Organisms within an ecosystem are organised into communities.

Organisation in an ecosystem

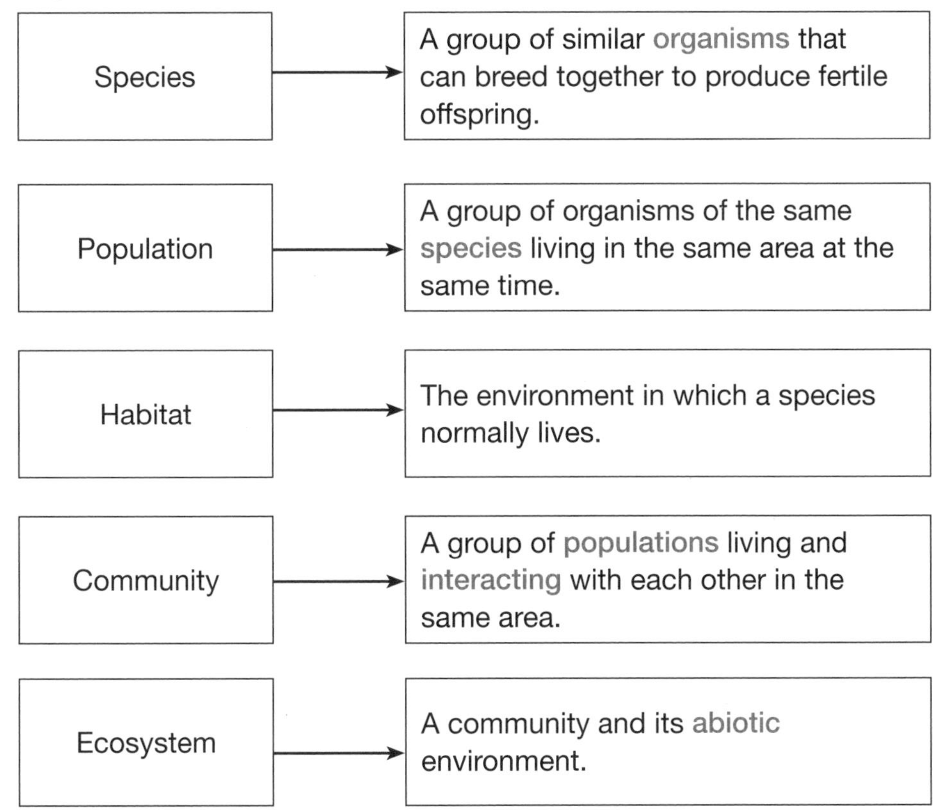

Each habitat has its own community, made up of populations of different species. These species interact with each other to compete for resources. All of the habitats make up the ecosystem.

Levels of organisation

Competition

Individuals in a community compete with others of the same species (intraspecific competition) and with individuals from other species (interspecific competition). Individuals compete for:

- food – animals
- water – animals and plants
- territory/space – animals and plants
- light – plants
- mineral ions – plants
- mates – animals.

Make a table of factors that species compete for, and whether the competition for these is intraspecific, interspecific or both.

Interdependence

Species depend on each other for survival. They depend on each other to provide food, shelter, pollination and seed dispersal. If one species is removed from a habitat, it can affect the whole community. This is called interdependence. For example, bees pollinate many flowering species. If the bee population decreases, the plants will not be pollinated and the plant species will also decrease.

Predator–prey relationships

In a stable community, the numbers of predators and prey rise and fall in cycles. An increase in the number of prey in a population provides more food for predators, so the number of predators in a population rises, after a slight delay. This in turn causes a decrease in the number of prey in a population and then a decrease in the number of predators in a population. This change in population sizes can be recorded over many years and displayed as a predator–prey cycle.

WORKIT!

This graph shows the interaction of toads and beetles. Describe and explain the shape of the graph. (5 marks)

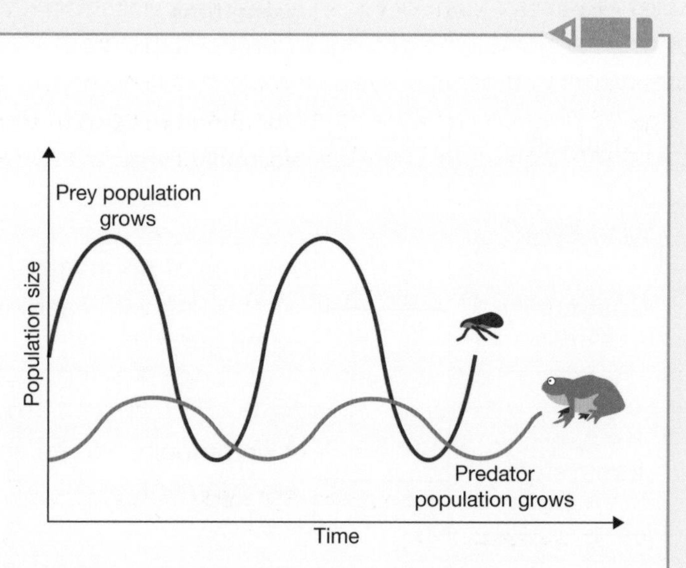

As the population of beetles increases, the population of toads increases. (1) This is because there is an increase in the food supply for the toads. (1)

As the population of beetles decreases, the population of toads decreases. (1) This is because there is a decrease in the food supply for toads. (1)

There is a delay between the increase or decrease of the beetle population and the increase or decrease of the toad population. (1)

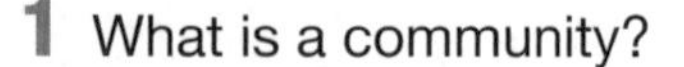

CHECKIT!

1 What is a community?

2 Two species of caterpillar are eating leaves in the same tree. What type of competition is this?

3 Mayfly nymphs are eaten by a species of frog that is in turn eaten by grass snakes. Suggest what would happen to the population of frogs and mayfly nymph if the grass snake population suddenly decreases.

Abiotic factors

An abiotic factor is a non-living condition that can affect where organisms live, e.g. temperature.

Make a table to show which of these abiotic factors would affect animals, plants or both.

Examples of abiotic factors

There are many abiotic factors that can affect a community. These include:

- temperature
- light intensity
- moisture levels in the soil and air
- pH of the soil.
- wind intensity and direction
- carbon dioxide levels
- oxygen levels of the water

How abiotic factors affect a community

Most organisms cannot live in extreme environments where it is very hot or cold, or there is no access to water. There are very few microscopic organisms that can survive in these conditions (see extremophiles on page 129). Even in stable communities, changes in the abiotic factors can affect the growth and number of organisms in a population.

WORKIT!

The light intensity in a woodland area was measured as well as the height of the flowers in that area. The results are shown in this graph. Describe and explain the heights of the flowers in different light intensities. (3 marks)

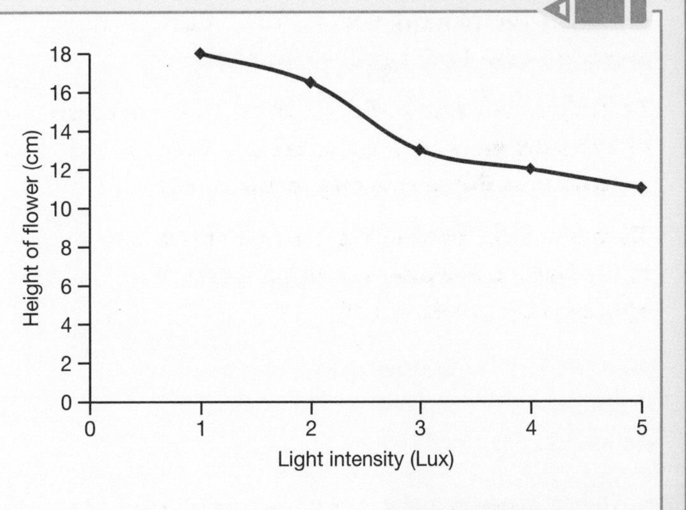

As the light intensity decreases, the height of the flowers increases. (1)

The height of the flowers increases from 11 cm to 13 cm between 5 to 3 lux, and from 13 cm to 18 cm between 3 and 1 lux. (1)

This is because the flowers are trying to get to the light in order to photosynthesise. (1)

Mention photosynthesis in your answer.

CHECKIT!

1 Give two examples of an abiotic factor.

2 Explain how the pH of the soil may affect organisms in a community.

3 Some trees live on a very windy hillside. Suggest how the wind may affect their growth.

Biotic factors

DO IT!

Write each biotic factor on a revision card and write a short description of how it can affect another organism or the environment.

STRETCH IT!

Research an example of symbiosis.

A biotic factor is any living component that affects the population of another organism or the environment.

Examples of biotic factors

Biotic factors affect the community. Some examples of biotic factors are:

food availability; predators; microorganisms in the soil; pathogens; competition within and between species; parasites; symbiosis; pollination.

How biotic factors affect the community

All organisms belong to a food chain or food web. The increase or decrease of one species will therefore directly or indirectly impact the population numbers of another species. Disease, predation, and parasites can also decrease a population's numbers, but symbiosis is beneficial to both species involved.

WORK IT!

Tarnished plant bugs (TPBs) are a pest that eats alfalfa plants in the USA. Parasites that kill TPBs were introduced to alfalfa fields in 1982. Use data from the graph to explain the effect of the parasites on the TPBs. (4 marks)

At first, as the percentage of parasitism increased, the number of TPBs remained constant. (1)

Around 1987-89, as the percentage of parasitism increased, the number of TPBs decreased. (1)

This is because the parasites killed some of the TPBs. (1)

After 1990-1992, the percentage of parasitism decreased as the number of TPBs had also decreased. (1)

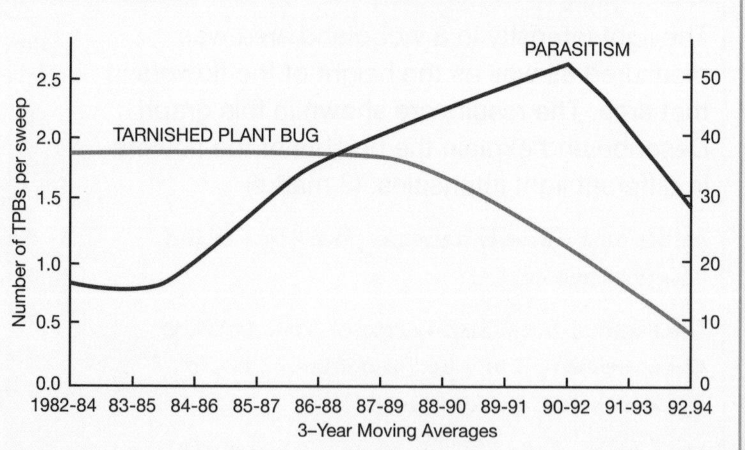

CHECK IT!

1 Give two examples of a biotic factor.

2 Describe the effect of the loss of a tree species on the community.

3 A community has a population of peas, caterpillars, bees and sparrows. A new pathogen has decreased the number of bees in the community. Suggest the effect that this will have on the other species.

Adaptations

All organisms are adapted to live in their environment.

Adaptations enable species to survive in the conditions in which they normally live, for example, a cold climate. There are three types of adaptation:

1 **Structural adaptations** are adaptations to the body of the organism, such as the skeleton, body shape, body colouration or fur length. For example, dolphins have streamlined bodies to aid swimming in the water.

2 **Behavioural adaptations** are changes to a species' behaviour to help their survival. Some examples are: emperor penguins huddle together for warmth during the Arctic winter; zebras stay together as a group when grazing so that their colouration confuses predators.

3 **A functional adaptation** is one that has occurred through natural selection over many generations in order to overcome a functional problem, such as birds having to eat seeds due to too much competition for other food sources, so the beak gradually becomes more adapted to seed eating.

Extremophiles

Some organisms are adapted to live in extreme environments, such as volcanoes or hot springs. These are usually microorganisms and are called extremophiles. Extremophiles can be found in habitats that have a high temperature, high pressure (under the sea) or a high salt concentration.

DO IT!

Make a table of examples of structural, behavioural and functional adaptations.

STRETCH IT!

Research extremophiles at deep-sea vents and find out how they are adapted to live at such high temperatures.

NAIL IT!

You will be expected to explain how a particular organism is adapted to its environment, given some information.

WORK IT!

Describe and explain how a camel is adapted to living in its environment. (4 marks)

Long eyelashes to keep the sand out of its eyes. (1) Large flat feet to support its weight on the sand. (1) Large fat store in hump to provide energy. (1) Does not sweat in high temperatures to keep water inside its body. (1)

CHECK IT!

1 Give an example of a structural adaptation in an Arctic animal.

2 What is the difference between a behavioural adaptation and a functional adaptation?

3 Explain how extremophiles can live in deep-sea hydrothermal vents.

Food chains

All organisms belong to a food chain or food web.

Photosynthetic organisms

The vast majority of food chains start with an organism that carries out photosynthesis. These organisms can be plants, algae, or bacteria that contain chlorophyll. These are called photosynthetic organisms. Light energy from the Sun is converted into biomass, which forms the basis of all food chains.

DO IT!

Draw an example of a food chain and expand it to form a food web.

Food chains

Feeding relationships within a community can be represented with food chains. Photosynthetic organisms are known as producers. These are eaten by primary consumers, which in turn are eaten by secondary consumers. Secondary consumers are eaten by tertiary consumers. Food chains link together to form food webs.

One example of a food chain is shown below. The sweetcorn is eaten by crickets, which are eaten by lizards, which are eaten by snakes. The arrow represents the transfer of energy from one organism to another.

The lizards and snakes are predators because they kill and eat other animals. The crickets and the lizards are prey because they are eaten.

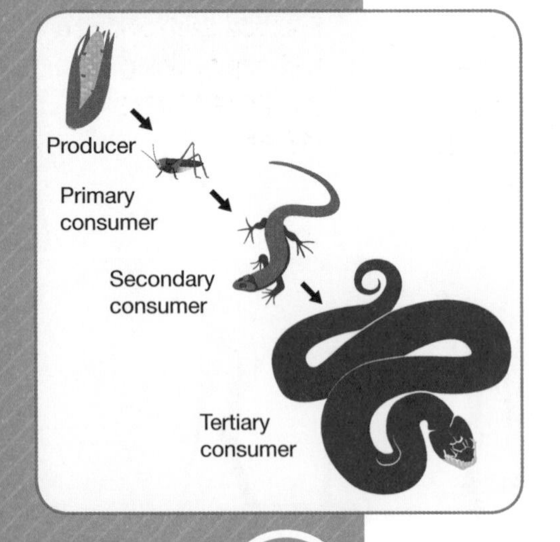

WORK IT!

Describe the energy flow through a food chain that consists of cabbages and caterpillars. (3 marks)

Light energy from the Sun is absorbed by the cabbages by photosynthesis. (1)

The cabbages transfer the light energy into biomass. (1)

The biomass of the cabbages is eaten by the caterpillars and converted into their own biomass. (1)

NAIL IT!

For Q3, think about the amount of energy being transferred.

CHECK IT!

1 What is a secondary consumer?

2 Draw and label a food chain to show the feeding relationship between grass, foxes and rabbits.

3 Suggest why there are often no more than four levels in a food chain.

Measuring species

Species are measured using a range of different sampling techniques.

Sampling techniques

It is difficult and time consuming to count every organism in a population. Sampling methods are quicker and use estimates of distribution and abundance of organisms in order to work out the size of the population.

Quadrats

Plants and slow-moving insects can be sampled using quadrats. This is a 1 m by 1 m square made of wire, which may be divided into smaller squares. The quadrat is placed on the ground and the number and type of each species is recorded. The quadrats are randomly placed in the area being investigated.

Transects

A transect is a long line made with string or a measuring tape, with quadrats placed at intervals, usually every metre, along it. The number and type of each species is recorded in each quadrat. Transects are usually used where the habitat changes over a short distance, for example, on a shoreline.

Animal traps

Animals can be trapped safely and then released again using a number of different traps, including nets, pooters and pitfall traps. Nets are used to sweep for larger insects. A pooter is a small jar with two straws and is useful for catching small insects. Suction is applied to one straw, so that the insect is sucked into the jar. A pitfall trap is a small beaker set into the ground with a raised lid to allow animals to climb in. These are useful for catching small ground invertebrates, such as woodlice.

DO IT!

Practice each of these techniques using equipment from your school.

Abundance of organisms

If you are using quadrats, the abundance of organisms is estimated by multiplying the average number of organisms found in the quadrats by the size of the area. The more quadrats used, the more reliable this estimate will be.

WORKIT!

24 daisies were found in 10 quadrats. The area measured 20 m by 20 m. Estimate the abundance of the daisies. (4 marks)

Average number of daisies per quadrat = 24 ÷ 10

= 2.4 (1)

Area is 20 m × 20 m = 400 m^2 (1)

Estimated number of daisies = 2.4 × 400 (1)

= 960 daisies (1)

If you are capturing and releasing animals, then you can use the capture, release, recapture method for measuring abundance. When you capture an animal, mark it in a safe way for the animal and release it. Then wait for a time and then capture animals in the same area again.

$$\text{Population size} = \frac{\text{number in first sample x number in second sample}}{\text{number in second sample that are marked}}$$

WORKIT!

18 beetles were captured in an area of woodland and marked. 20 beetles were caught the second time, eight of these were marked. Estimate the abundance of the beetles. (3 marks)

$$\text{Abundance of beetles} = \frac{18 \times 20}{8}$$

$$= \frac{360}{8}$$

= 45 beetles (3)

1 Describe how to use a quadrat.

2 If five quadrats are used in a 100 m^2 area, and measure 25 buttercups, what is the abundance of buttercups in the area?

3 Explain why it is important to mark captured animals in a safe way.

Required practical 9: Measuring the population size of a common species

In this practical you will use sampling techniques to investigate the effect of a factor on the distribution of a common species.

Practical Skills

Investigation

Choose a local area in which to carry out your investigation. Use quadrats, a transect or animal traps to sample a common species.

Recording your findings

Record your results in a table. If using quadrats or a transect, record the number of individuals of a species in each quadrat, and then find the mean number of species per quadrat.

If using animal traps to count individuals of a species, mark your captured animals and release them. Then wait a period of time and capture animals in the same area.

Estimating population size

To estimate the population size of plants, you need to multiply the mean number of plants per quadrat by the size of the investigation area.

WORKIT!

In a woodland area measuring 100 m^2, there are 14 bluebells per quadrat. Estimate the population size of the bluebells. (2 marks)

Population size = $14 \times 100\ m^2$

= 1400 bluebells

WORKIT!

140 bluebells were found in 10 quadrats in an area of woodland. What is the mean number of bluebells per quadrat? (1 mark)

$140 \div 10 = 14$ bluebells per quadrat. (1)

NAILIT!

To estimate the population size of animals, you need to know the number of individuals you captured each time, and how many were marked.

Assumptions of capture–release–recapture

When estimating the population size of animals using this method, you must assume: no animals die or are born, no animals migrate and the marking of the animals has not affected their survival.

1 Describe how to use an animal trap to estimate the population of an animal species.

2 If 12 snails are caught and marked, 11 are recaptured and 8 are marked, estimate the population size.

3 Explain why it must be assumed that there are no deaths, births or migration in a population when you estimate the population size.

The carbon cycle

Carbon is cycled through the ecosystem by biological and chemical processes.

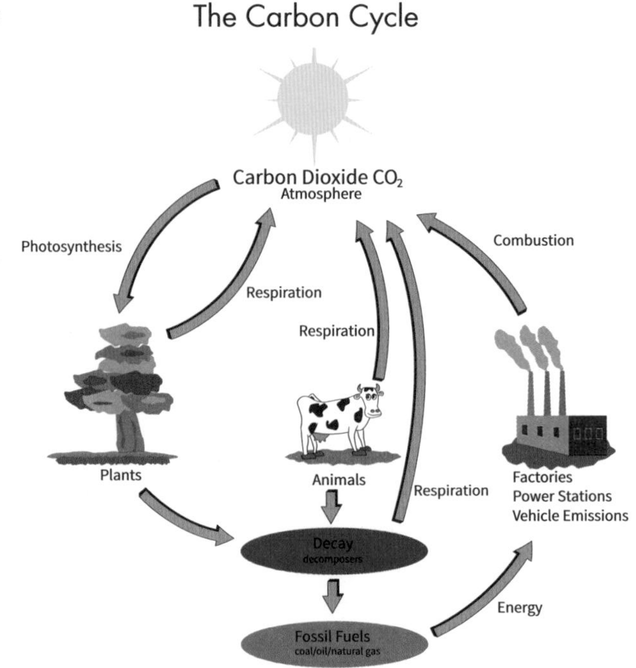

How carbon is cycled through the ecosystem

Carbon dioxide in the air is absorbed by plants to use in photosynthesis. Plants convert carbon dioxide into sugars and the carbon becomes part of the biomass of the plant. When consumers eat plants, this carbon is passed to them. When plants and animals die their remains decay, and are decomposed by microorganisms, such as fungi and bacteria. These are known as the decomposers. Millions of years ago, some of their remains formed fossil fuels, such as coal or oil.

Carbon passes back into the atmosphere as carbon dioxide by:

- aerobic respiration of plants, animals and microorganisms
- combustion – the burning of fossil fuels.

SNAPIT!

Make your own copy of the Carbon Cycle and write notes for each arrow explaining in more detail how the carbon is being transferred. Take a photo of your diagram.

WORKIT!

Explain the importance of the carbon cycle to living organisms. (4 marks)

Carbon dioxide is needed by plants in order to carry out photosynthesis. (1)

Photosynthesis produces sugars, which are used in respiration to provide energy for the plant. (1)

Sugars are eaten by animals and are used in respiration to provide energy for the animal. (1)

When dead matter is decomposed, the decomposers use the sugars in respiration to provide energy. (1)

CHECKIT!

1 Name one way in which carbon dioxide returns to the atmosphere.

2 Explain why microorganisms are important in the carbon cycle.

3 Explain why there are increasing levels of carbon dioxide in the atmosphere.

The water cycle

Water is cycled through the environment through the processes of evaporation and precipitation (rain).

Water in the oceans is warmed by the Sun, and evaporates into the atmosphere. The water droplets condense to form clouds and are transported inland by the wind. As the clouds rise over mountains, the water droplets are released as precipitation. The water moves under the land as ground water, and over the land into streams and rivers, and finally makes its way back to the oceans.

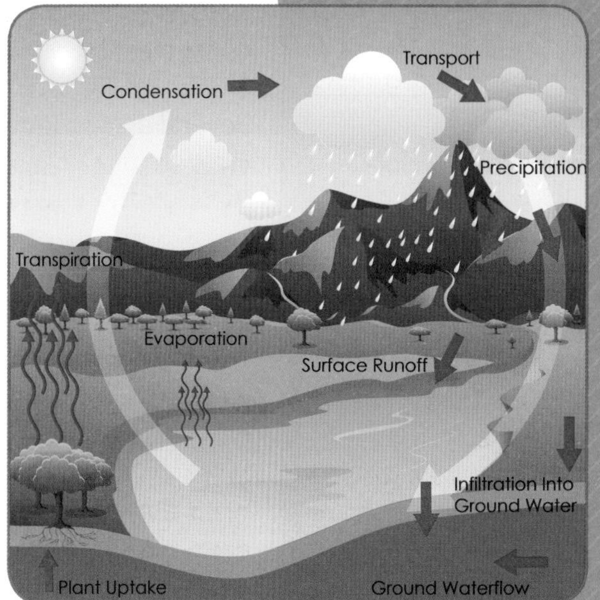

The Water Cycle

The importance of the water cycle

The water cycle is important to living organisms because it provides fresh water for plants and animals on land. Fresh water is needed for animals to drink, and for plants to use in photosynthesis. Water diffuses from the top surfaces of plants in a process called transpiration (see page 45).

WORKIT!

Describe and explain the role of plants in the water cycle. (3 marks)

Plants take up fresh water for use in photosynthesis. (1)

Water transpires from the surface of the plant. (1)

Water droplets return to the atmosphere and condense into clouds. (1)

SNAPIT!

Make your own copy of the water cycle diagram, including the key words and their definitions. Take a photo to revise from later.

CHECKIT!

1 Name two ways in which water can return to the atmosphere.

2 Explain why water evaporates form the surface of the oceans.

3 The land on one side of a mountain receives a large amount of rain, whereas the land on the other side of the mountain receives very little rain. Suggest how this is possible.

Decomposition

When organisms die, their remains decay, or decompose, through the action of decomposers.

DO IT!

Write down each of these abiotic factors onto revision cards and learn them.

The influence of temperature, water and oxygen

Several abiotic factors influence the rate of decomposition:

Temperature – High temperatures denature enzymes and proteins, and kill the decomposers. This prevents decay.

– Low temperatures slow down the rate of reaction of enzymes and so slow down the rate of decay.

Water – Decomposers need water for processes within their bodies, such as transport.

– A lack of water will slow down or prevent the rate of decay.

Oxygen – Decomposers need oxygen for aerobic respiration.

– A lack of oxygen will slow down or prevent the rate of decay.

Increasing the rate of decay

Gardeners and farmers try to increase the rate of decay of waste biological material by providing the optimum conditions. They also introduce detritivores, such as worms and woodlice to break down the decaying matter into smaller pieces. This speeds up the production of compost, which can be used as a natural fertiliser for growing crops or garden plants.

Waste biological material can also be placed inside biogas generators, where the material decays anaerobically. Anaerobic microorganisms decompose the waste material to produce methane gas. This can then be used as a fuel.

WORK IT!

Some decaying matter was left for 4 days. The initial mass was 500 g and the final mass was 300 g. What is the rate of decay? (2 marks)

$$\text{Rate of decay} = \frac{500\,g - 300\,g}{4}$$

$$= 200\,g/4$$

$$= 50\,g/day \quad (2)$$

Calculating the rate of decay

In order to calculate the rate of decay of compost, you need to know the initial mass of the decaying matter, the final mass of the decaying matter and the number of days it has been decaying for.

$$\text{Rate of decay} = \frac{\text{initial mass (g)} - \text{final mass (g)}}{\text{number of days}}$$

CHECK IT!

1. Give one way in which gardeners or farmers could increase the rate of decay.
2. Calculate the rate of decay for some decaying matter that took 5 days to decay from 1 000 g to 600 g.
3. A food company wants to transport tomatoes from France to the UK. What could the company do to prevent the tomatoes from decaying?

Required practical 10: Investigating the effect of temperature in the rate of decay

In this practical you will need to hypothesize about the effect of temperature on the rate of decay of a biological molecule. Temperature affects the enzymes within decomposers and so affects the rate of decay.

Practical Skills

The effect of **temperature** on the rate of decay of milk

- Add a pH indicator, usually **phenolphthalein**, to the milk.
- Time how long it would take for the **pH** to change by timing how long it took the phenolphthalein to change colour, from pink to colourless.
- Repeat your investigation at different temperatures. To warm the water, either use a water bath, or a beaker of water on a tripod over a Bunsen burner.
- Carry out your investigation at least three times, in order to make your results **reliable**. Keep all of your variables the same, except for the **independent variable**, to increase the **validity** of your results.
- Record your results in a table and find the mean time for each temperature and calculate the rate of reaction for each temperature.

NAILIT!

Phenolphthalein is pink in alkaline solutions (pH8.2 or above) and colourless in acidic or near neutral solutions (pH8.1 or below).

DOIT!

Write out your method, step by step. Write down the equipment you used, the volumes of any reactants and which were your dependent and independent variables.

WORKIT!

This table shows the results of an investigation into the decay of milk at different temperatures. Complete the table. (4 marks)

Temperature (°C)	Time taken for pH to change (s)			Mean time taken for pH to change (s)	Rate of reaction
	1	2	3		
20	230	218	242	230	0.004
40	121	132	110	121	0.008
60	260	232	256	249	0.004
80	410	421	397	409	0.002

(4)

NAILIT!

Remember to consider which variables will need to stay the same (control variables) and to evaluate your method, identifying possible improvements.

CHECKIT!

1 Using the table above, describe the effect of temperature on the rate of decay of milk.

2 In this investigation, what is the dependent variable?

3 Explain why the pH of the milk changes during decay.

Impact of environmental change

Environmental changes affect the distribution of species in an ecosystem.

Temperature

Global warming, due to increased carbon dioxide in the atmosphere, increases the average global temperature. The temperature of a climate affects the plants that can grow there. This affects the amount and type of food available. A high temperature can cause desertification.

Water availability

As global temperatures rise, water availability decreases. All organisms need water and if there is not enough water, then many species will not survive.

Atmospheric gases

Deforestation increases the amount of land for houses and farming. However, the decrease in the number of trees decreases the amount of photosynthesis and this increases the amount of carbon dioxide in the atmosphere. Human activity, such as burning fossil fuels, also increases the amount of carbon dioxide in the atmosphere.

Species will migrate to an area where the temperature is more suitable, and there is available water. Sometimes the changes in temperature and water availability are seasonal, and species migrate between geographical areas for a season.

Biodiversity

The higher the level of biodiversity, the more stable the ecosystem is. This is because a species can rely on several other species for food and shelter.

Our future depends on maintaining a high level of biodiversity on the Earth. Human activities, such as pollution, burning fossil fuels and deforestation are decreasing the level of biodiversity.

DO IT!

Make revision cards of each abiotic factor that affects the distribution of species.

NAIL IT!

An abiotic factor is a non-living part of the environment that affects living organisms.

STRETCH IT!

Find out about peat bog habitats and the species that live there.

WORK IT!

Explain how waste, deforestation and global warming have an impact on biodiversity. (3 marks)

Waste pollutes the water and air and decreases the number of organisms in an ecosystem. (1) Deforestation reduces photosynthesis and increases the levels of carbon dioxide in the atmosphere. (1) This increases global warming, which increases the temperature of an ecosystem, decreasing the number of organisms that live there. (1)

CHECK IT!

1. Name an abiotic factor that has an effect on the distribution of species.
2. Describe how global warming can cause desertification.
3. Suggest what could happen to an animal species if its habitat increased in temperature.

Biodiversity

Biodiversity is the variety of living organisms on the Earth, or within an ecosystem.

Waste management

The rapid growth of the human population and increasing standard of living, means that humans are using more and more resources. This means that more waste and chemical materials are produced, which in turn leads to increased pollution:

- in water – sewage, fertiliser leaching off from fields, toxic chemicals, e.g. oil spills
- in air – smoke and acidic gases from vehicle exhausts and power stations
- on land – decomposition of landfill and from toxic chemicals, e.g. pesticides.

Increasing levels of pollution lead to a decrease in biodiversity.

Land use

When humans clear land to build houses, or quarry for stone, or grow crops, or dump waste, it leaves fewer habitats for other organisms. This leads to decreased biodiversity.

Peat bogs have been dug up for many years to provide peat for compost. However, this destroys the beat bog habitat and decreases the biodiversity of the species that live there, including plants, animals and microorganisms. Burning peat also causes large amounts of **carbon dioxide** to be released into the atmosphere.

Make revision cards of each type of pollution.

Find out about some programs from organisations, such as the WWF, that are protecting species from extinction.

WORKIT!

Explain why peat bogs need to be protected as the human population and its food production increases.

As the global population increases, more food needs to be grown. (1)

Peat from peat bogs is a cheap and effective compost and increases the amount of food that can be produced. (1)

Peat bogs should be protected as destroying them decreases the biodiversity of the habitat. (1)

CHECKIT!

1. What type of pollution do power stations produce?
2. Describe how using land for building houses decreases biodiversity.
3. Describe and explain the role of peat bogs in the carbon cycle.

Global warming

Global warming is the gradual increase in the overall temperature of the Earth's atmosphere.

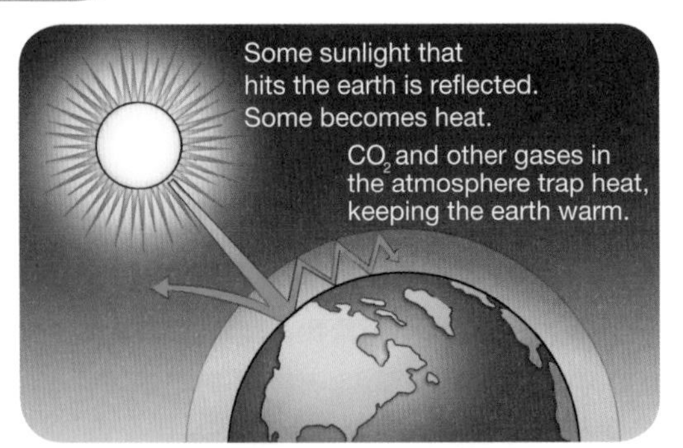

The Greenhouse Effect

Causes of global warming

Global warming is caused by the greenhouse effect. This is when carbon dioxide, methane and other greenhouse gases build up in the Earth's atmosphere. Radiation from the Sun warms the Earth, and this heat is reflected from the Earth's surface. The greenhouses gases in the atmosphere absorb the heat, increasing the temperature of the atmosphere.

Greenhouse gases are increasing. They build up due to the respiration of organisms and the combustion of fossil fuels.

SNAPIT!

Draw your own version of this diagram and learn the process of global warming. Take a photo to help you revise later.

Biological consequences of global warming

Global warming has many consequences for the organisms on Earth:

- global weather patterns will change – causing flooding in some areas and drought in other areas. This will decrease available habitats, and food and water availability
- sea levels will rise – decreasing available habitats
- increased migration of organisms – species will move to more suitable habitats with enough available water and food
- increased extinction of species – some species will not be able to migrate or adapt quickly enough to the changing climate.

DOIT!

Make a revision card for each of the biological consequences.

WORKIT!

A species of mountain beetle has been found living higher up the mountain in recent years. Explain why this could have happened. (4 marks)

The mountain beetles feed on plants that grow at certain heights on the mountain. (1)

The plants have an optimum temperature that they like to grow in. (1)

As the temperature on the mountain increases, the plants cannot survive in the warmer areas. (1)

The plants begin to grow higher up the mountain, where it is cooler, and the beetles migrate to feed on them. (1)

CHECKIT!

1. Name two greenhouse gases.
2. Describe the process of global warming.
3. Describe and explain human activities that could decrease the amount of greenhouse gases in the atmosphere.

Maintaining biodiversity

There are both positive and negative human interactions in an ecosystem that can impact biodiversity.

Positive interactions

- Conservation programmes – protecting species from extinction.
- Using renewable energy – reducing the amounts of greenhouses gases in the atmosphere and reducing the effects of global warming.
- Recycling waste – reusing resources to prevent pollution of the environment by dumping waste.

Negative interactions

- Combustion of fossil fuels – increases the amounts of greenhouse gases in the atmosphere, and increases the effect of global warming.
- Pollution of the air, water and land – reduces the biodiversity of the polluted areas.
- Deforestation – removes habitats for organisms, and decreases photosynthesis, which contributes to global warming.

Programmes to maintain biodiversity

Concerned people have put many programmes into place in order to reduce the negative effects of humans on ecosystems and biodiversity. These include:

Breeding programmes for endangered species, e.g. in zoos or wildlife reserves; protection and regeneration of rare habitats, e.g. conservation areas, to which people have limited access; reintroduction of hedgerows around fields where farmers only grow one type of crop; reduction of deforestation and carbon dioxide emissions.

DO IT!

Make a table of the positive and negative interaction of humans with an ecosystem.

WORK IT!

Explain and evaluate the conflicting pressures on maintaining biodiversity in a rare habitat where people live. (3 marks)

People need land to live on and grow food, which reduces the land available for other organisms. (1) People create waste which needs to be removed, otherwise it will pollute the land and make it unsuitable for other organisms. (1) If rare habitats are not protected, many species will become extinct. (1)

CHECK IT!

1. Give one negative interaction that humans have with an ecosystem.
2. Explain how reintroducing hedgerows increases biodiversity.
3. The Galapagos Islands have many unique species and are visited by thousands of tourists each year. Suggest some ways that the impact of tourism to these islands can be reduced.

Trophic levels

DO IT!

Make revision cards of each trophic level and learn the keywords.

The trophic level of an organism is the position it occurs in the food chain.

The trophic levels are:

Producer – usually plants, they convert light energy from the Sun into chemical energy, through the process of photosynthesis.

Primary consumer – herbivores that consume the producers.

Secondary consumer – carnivores that consume the primary consumers.

Tertiary consumer – carnivores that eat other carnivores. Also called apex predators, as they have no predators of their own.

At each trophic level, when the organisms die, they are decomposed by microorganisms called decomposers. The decomposers secrete enzymes onto the decaying matter and then absorb the resulting food molecules into themselves.

SNAP IT!

Make up your own diagrams of the pyramids of biomass and practise drawing them for different food chains. Take a photo of what you have drawn.

Pyramids of biomass

The relative amount of biomass at each trophic level can be represented with a pyramid of biomass. Trophic level 1 is at the bottom of the pyramid.

starling
lacewings
aphids
elder tree

WORK IT!

In a food web, grass and cabbages are eaten by caterpillars and rabbits. The caterpillars are eaten by blackbirds, and the rabbits are eaten by foxes. The blackbirds and rabbits are eaten by eagles. Describe and explain the trophic level of the eagle. (3 marks)

The eagle is a secondary consumer (1) as it eats the rabbit, which is a primary consumer/herbivore. (1) The eagle is also a tertiary consumer (1) as it eats a blackbird which is a secondary consumer. (1)

CHECK IT!

1 Name the bottom layer of a pyramid of biomass.

2 Describe the role of secondary consumers.

3 Describe how decomposers get their energy.

Pyramids of biomass

The relative amount of biomass at each trophic level can be represented with a pyramid of biomass.

How to draw a pyramid of biomass

Each trophic level of the pyramid must be drawn to the same scale. For example, if the biomass of each trophic levels are 10 kg, 5 kg, 1 kg, then each kg could be represented by 1 cm. So the trophic levels of the pyramid would be drawn:

Secondary consumers (1 kg) – 1 cm
Primary consumers (5 kg) – 5 cm
Producers (10 kg) – 10 cm

Secondary consumers
Primary consumers
Producers

Transfer of biomass

Only 1% of the incidence light from the Sun is used in photosynthesis. Much of the rest of the light is reflected, passes through the leaves, or does not hit the leaves at all.

Only 10% of the biomass from each trophic level is passed to next level. The rest is lost due to:

- egestion of non-digestible material as faeces
- waste, such as exhalation of carbon dioxide and water from respiration
- loss of water and urea in urine.

Large amounts of glucose are used in respiration, and do not form the biomass of the organism. At each level of the food chain, the biomass gets smaller. This is why there are not usually more than four or five trophic levels in a food chain.

Calculating the efficiency of biomass transfers

You need to be able to calculate the efficiency of biomass transfers between trophic levels using percentages or fractions of mass.

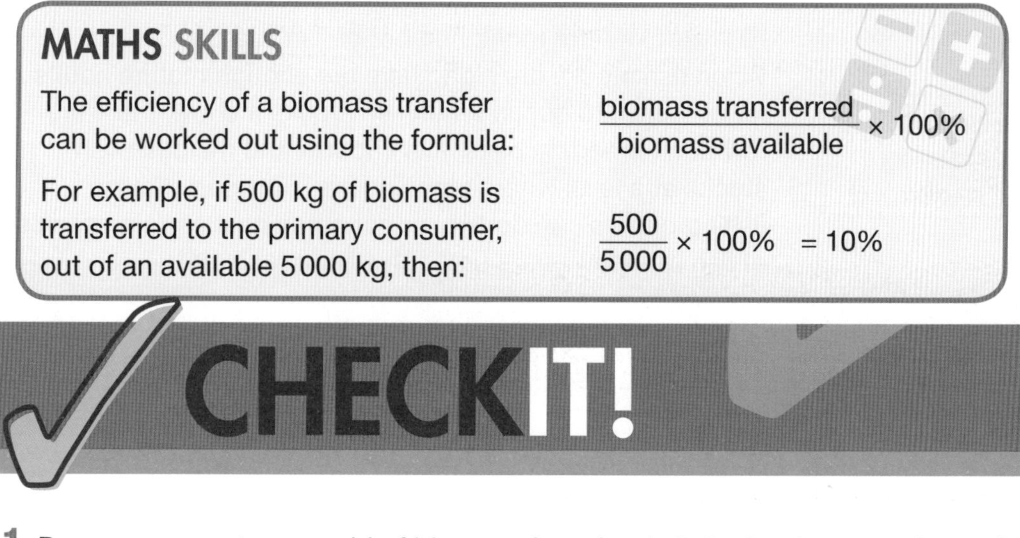

MATHS SKILLS

The efficiency of a biomass transfer can be worked out using the formula:

$$\frac{\text{biomass transferred}}{\text{biomass available}} \times 100\%$$

For example, if 500 kg of biomass is transferred to the primary consumer, out of an available 5 000 kg, then:

$$\frac{500}{5000} \times 100\% = 10\%$$

DO IT!

Practise drawing pyramids of biomass using food chains. Remember that the height of each level must remain constant.

NAIL IT!

Always use a sharp pencil and a ruler when drawing pyramids of biomass.

WORK IT!

Explain how the efficiency of biomass transfers affects the number of organisms at each trophic level. (3 marks)

Only 10% of the biomass from each trophic level gets passed to the next trophic level. (1)

Therefore, the biomass of the trophic level gets smaller as you go up the food chain. (1)

This means that there are fewer organisms at each trophic level as you go up the food chain. (1)

CHECK IT!

1. Draw an accurate pyramid of biomass for a food chain that has an oak tree (120 kg), caterpillars (5 kg) and blackbirds (2 kg).
2. Calculate the efficiency of biomass transfer, if there is 1 200 kg of biomass available, but only 400 kg is transferred.
3. Explain why 90% of the biomass is not transferred to the next trophic level.

Food production

Food security means having enough food to feed a population.

Biological factors affecting food security

There are many biological factors that affect food security:

- increasing birth rate in some countries limits the amount of food and water available
- changing diets in developed countries means that scarce food sources are transported around the world
- new pests and pathogens that affect farming
- environmental changes affect food production, such as rains failing causing widespread famine
- the cost of agricultural inputs, such as fertilisers and pesticides
- conflicts in some parts of the world that affect the availability of water or food.

Sustainable methods must be found to feed all of the people on Earth.

Make revision cards of each biological factor explaining how it reduces water or food availability.

Farming techniques

Restricting the loss of energy from food animals to the environment can increase the efficiency of food production. This can be done by limiting the animals' movement by keeping them in cages or barns, and controlling the temperature of their surroundings.

Some animals are fed high-protein foods to increase growth.

Some people have ethical objections to some modern intensive farming techniques, such as keeping animals in confined spaces.

WORKIT!

Evaluate the advantages and disadvantages of intensively farming animals. (3 marks)

Food animals do not lose as much energy because their movement is restricted. (1)

This increases the biomass of the farm animals and provides more food for people. (1)

Some people think that keeping animals in confined spaces is unethical. (1)

Intensively farmed animals are more likely to suffer from diseases, as diseases can more easily spread from one animal to another. (1)

Sustainable fisheries

Fish stocks refer to the number of fish in a population. When fish stocks are high, there are enough mature adults for breeding to occur. When fish stocks are low, there are few adult breeding pairs. It is important to maintain fish stocks at a level where breeding can continue. Having fishing quotas can do this, or not catching a certain species of fish for a few years until fish stocks recover.

CHECKIT!

1. Name a biological factor that affects food security.
2. A country has banned the fishing of cod until the number of breeding adults reaches a certain number. Explain how this is a sustainable fishing technique.
3. A country of 50 million people has suffered a drought and only 50% of the crops have been harvested. Suggest what could be done to ensure the food security of the country in the future.

Role of biotechnology

Biotechnology is the use of microorganisms or animals to make a product.

In order to feed a growing global population, biotechnology can be used to produce larger quantities of food. For example, plants can be genetically modified to be resistant to:

- drought – can survive with less water
- pests – plants contain a pesticide that kills pests that try to eat them
- herbicides – herbicides used on crops will only kill weeds, not the crop itself.

Biotechnology can also be used to culture microorganisms for food. For example, the fungus *Fusarium* is used to make mycoprotein, a type of protein-rich fungus used as a meat substitute in vegetarian food. The fungus is grown in a glucose containing broth inside large fermenters. A large stirrer inside the fermenter mixes oxygen through the broth. The fungus is harvested from the fermenter and purified before being made into food products.

In some areas of the world, people do not get enough vitamins or minerals. Plants that are genetically modified to contain additional vitamins or minerals can be used to prevent deficiencies. For example, golden rice is a type of genetically modified rice that contains beta-carotene, which is needed to make vitamin A.

Using microorganisms to make insulin

Biotechnology can be used in medicine to make human proteins from animals or microorganisms. For example, transgenic goats can produce human blood clotting factors in their milk.

Human insulin is made in large fermenters using bacteria that have been genetically modified with the gene for human insulin. As the bacteria grow, they produce human insulin. This can be harvested and purified from the fermenters for use by people with diabetes.

DO IT!

Watch an online video on how mycoprotein is made.

STRETCH IT!

Find out about some other human proteins made by genetically modified animals or microorganisms.

WORK IT!

Explain how biotechnology could help to improve food security in the future. (3 marks)

Genetically modified crops could produce a higher yield of crops. (1)

Crops can be genetically modified to resist drought. (1)

Mycoprotein can be grown as an alternative to meat. (1)

1. Give one advantage of a genetically modified plant.
2. Explain why fermenters need to have oxygen mixed through the broth.
3. Explain how genetically modified rice can be used to treat vitamin A deficiencies in some parts of the world.

REVIEW IT!

Ecology

1 a Define the term 'population'.

b What is a community?

c Give two examples of an abiotic factor.

2 a How is carbon dioxide in the air used in plants?

b Name two ways in which carbon dioxide is returned to the atmosphere.

c i Describe how decomposers contribute to the carbon cycle.

ii Give one abiotic factor that increases the rate of decomposition.

iii Bacteria cause milk to go sour by releasing lactic acid. It took 240 seconds for the pH to change. Calculate the rate of decay.

3 a Describe how global warming has affected:

i global temperatures

ii water availability

iii atmospheric gases.

b What are the biological consequences of global warming?

4 a Name two biological factors that affect food security.

b Explain how farmers can increase the efficiency of food production in livestock.

5 An aquatic food chain is shown below:

plankton → shrimp → herring → shark

70 kg 30 kg 20 kg 15 kg

a Which organism is the producer?

b Which organism is the secondary consumer?

c Draw a pyramid of biomass for the food chain.

d Explain what would happen to the number of sharks if the number of shrimp decreased?

6 a Some students were estimating the number of buttercups growing in the 10 m by 15 m school field. They used 10 quadrats and recorded their data in the table below:

	Quadrats										Total number of buttercups
	1	2	3	4	5	6	7	8	9	10	
Number of buttercups	1	2	0	10	4	6	2	1	6	2	

i Complete the table.

ii Use data from the table to estimate the abundance of the buttercups in the field.

b The students then decided to estimate the abundance of snails, using the capture, release, recapture method. They caught and marked 25 snails the first time, and caught 18 snails the second time. Of these, 10 were marked. Estimate the abundance of the snails.

Glossary / Index

Page references given in bold

A

Abiotic An abiotic factor is a non-living condition that can affect where organisms live, e.g. temperature. **127**

Absorb The process of absorbing substances into cells or across the tissues and organs through diffusion or osmosis. **28, 71, 86-88**

Abundance The number of individuals of each species in a sample. **131, 132, 146**

Acidic gases Gases in the atmosphere that can combine with rain water to produce acid rain. **139**

Adaptation Adaptations enable species to survive in the conditions in which they normally live, for example, a cold climate. **129**

Adhesion The attraction between water molecules and the xylem wall in transpiration. **45**

Adrenal gland The gland above the kidney that secretes hormones such as adrenaline. **82, 94**

Adrenaline The hormone that increases heart rate and breathing rate when a peron is scared or excited. **94**

Adult cell cloning A method of cloning using DNA from an adult cell and an empty egg cell. **115**

Aerobic respiration The process of using oxygen to break down glucose to produce energy, making carbon dioxide and water as byproducts. **68, 134, 136**

Alfred Russel Wallace The biologist that came up with the theory of evolution at the same time as Charles Darwin. **118**

Alleles A version of a gene. **105, 107–9, 114, 119**

Allopatric speciation A form of speciation when the two populations are geographically separated. **118**

Alveoli Small air sacs in the lungs that are the site of gaseous exchange. **22, 35**

Amino acids The small units, or monomers, that proteins are made from. **87, 102, 103**

Ammonia A toxic chemical that is produced as a product of amino acid deamination. **87**

Amylase An enzyme that breaks down starch. **29, 30, 33**

Anaerobic respiration The process of breaking down glucose to produce energy in the absence of oxygen, making carbon dioxide and lactic acid as byproducts. **68–9, 70**

Angina Chest pains, often brought on by exercise, as the blood supply to the muscles of the heart is restricted. **38**

Antibacterial chemicals Chemicals that kill bacteria. **61**

Antibiotic resistance When bacteria cannot be killed by some or all antibiotics. **51, 55, 120–1**

Antibiotics Medicines that kill bacteria, or slow down their growth. **17, 51, 55, 120–1**

Antibody A protein that binds to a specific antigen on a pathogen. **53, 54, 57–8**

Antigen A foreign substance that triggers an immune response in the body. **57**

Antimalarial drugs A medicine that kills the protozoa that causes malaria. **52**

Antiseptic A substance which inhibits the growth and development of microorganisms. **17**

Antitoxins Antibodies that bind to the toxins produced by microorganisms in the body. **53**

Antiviral Medicines that kill virusues. **55**

Aorta The artery that carries oxygenated blood away from the heart and the largest artery in the body. **34**

Arteriole A small artery. **86**

Artificial heart A mechanical heart that can be used in transplants to aid or replace the heart. **39**

Artificial pacemaker A small mechanical device that coordinates the resting heartbeat. **34**

Artificial selection Selective breeding of organisms to produce offspring with the desired characteristics. **112**

Aseptic technique A procedure that is performed under sterile conditions. **13**

Asexual reproduction A form of reproduction where the offspring are clones of the parent. **98–9**

Automatic control Processes in the body, controlled by the brain, that are involuntary. **73, 75**

Auxin A plant hormone that controls the growth of shoots towards the light and roots towards gravity. **95, 115**

Axon The long thin section of a neurone along which the nerve impulses travel. **10, 75, 76**

B

Bacteria Unicellular, prokaryotic microorganisms. **8, 22**

Barrier method A method of preventing sperm form reaching an egg during sexual intercourse e.g. condoms. **92**

Behavioural adaptation Changes to a species' behaviour to help their survival e.g. penguins huddling together for warmth. **129**

Benign tumour A growth of abnormal cells, contained in one area, that does not invade other parts of the body. **43**

Bias A conclusion that may be incorrect. **56**

Bile A substance produced by the liver that emulsifies fats into smaller droplets. **29**

Binary fission Asexual reproduction where bacteria divide their genetic material and double their normal size to make two daughter cells. **98**

Binding site The part of an antibody that binds in a complementary way to an antigen. **57**

Biodiversity The variety of living organisms on the Earth, or within an ecosystem. **138, 139, 141**

Biogas generator A machine that uses microorganisms to break down organic waste anaerobically to produce methane gas. **136**

Biomass The total mass of the individuals of a species in a given area. **130, 134, 142, 143**

Biotechnology The use of microorganisms or animals to make a product. **145**

Biotic Any living component that affects the population of another organism or the environment. **128**

Birth rate The number of offspring born in a year. **133, 144**

Bladder Organ that stores urine. **86**

Bowman's capsule Part of the nephron where small molecules are sieved from the glomerulus. **86**

Brain Organ that controls activity of the body. **73, 78, 81**

Breeding programmes A programme of breeding organisms together to produce more offspring. **141**

Bronchi The two branches from the trachea that lead into the lungs. **35**

Budding Asexual reproduction where yeast double their genetic material and organelles into small buds on their surface, which break off. **98**

C

Capillary Small blood vessels that carry blood around the body's tissues. **35, 36, 86**

Capture, release, recapture method A method of estimating population sizes by capturing organisms, marking them, releasing them and capturing some of them again. **132**

Carbon dioxide The gas produced by the body by respiration and used by plants in photosynthesis. **35, 63, 64, 69, 134**

Carrier molecules A molecule that carries other molecules to a given location. **103**

Carrier protein A protein embedded in a membrane that uses energy to open and close a channel through which large molecules can pass. **26**

Cell bodies Part of the neurone that contains the nucleus. **75**
Cell cycle The process of division in a cell that duplicates the organelles and the genetic material. **18**
Cellular respiration The respiration of the cell. **71**
Cellulose cell walls The cell walls of plant cells. **61**
Central nervous system The neurones in the brain and spinal cord. **74**
Cerebellum The part of the brain that controls balance and coordination. **78**
Cerebral cortex The part of the brain that controls conscious thought, memory, language and learning. **78**
Charles Darwin Biologist that came up with the theory of evolution by natural selection. **117, 118**
Chemical energy The energy found in chemicals and biological molecules e.g. food. **63, 68**
Chlorophyll The green pigment in chloroplasts. **60, 63, 130**
Chloroplast Organelle in plant cells that is the site of photosynthesis. **63**
Chlorosis The yellowing of the leaves, caused by disease, lack of sunlight, or mineral ion deficiency. **60**
Chromosomes A large coiled DNA molecule containing many genes. **18, 100, 101, 105, 109, 119**
Climate The weather conditions in an area in general or over a long period. **138, 140**
Clone Offspring that is genetically identical to the parent, or to another offspring. **19–20, 57, 98, 115–16**
Coding DNA DNA that contains the instructions to make a protein. **104**
Cohesion The attraction between water molecules in the water column in the xylem. **45**
Collecting duct Part of the kidney that reabsorbs water and collects urine. **88**
Combustion Burning of any flammable material. **134, 141**
Communities A group of populations living and interacting with each other in the same area. **125–8**
Competition When individuals in a population fight over resources. **125**
Complementary base pairing The pairing of DNA bases, A with T and C with G. **102**
Compost Decayed organic material which is used as a fertiliser for growing plants. **136, 139**
Concentration gradient The difference in the concentration of particles on either side of a partially permeable membrane. **26, 89**
Condensation When water vapour forms a liquid. **135**
Condom A barrier method worn by a man to prevent sperm from reaching an egg during sexual intercourse. **51, 92**
Conjugation When a bacterium passes on some genetic material to another bacterium. **120**
Conservation programmes Protection of species in their natural habitat or in zoos or botanical gardens to prevent extinction. **141**
Contraception A method of preventing pregnancy. **92**
Contract When the muscle fibres move together to create a high pressure or force. **10, 34, 70, 81**
Coordinator Usually the brain, controls what will happen in response to the stimulus. **74**
Cornea Area at the front of the eye that bends light towards the pupil. **79**
Coronary arteries Arteries that supply the heart muscle with blood. **34, 38–9**
Crenate When an animal cells loses too much water and shrivels. **23, 86**
Cutting A small section of plant used to make a new plant. **115**

D

Deaminated When amino acids are broken down into ammonia. **87**
Decayed When dead organisms or matter is decomposed. **59, 120, 134, 136, 137, 142, 146**
Decomposers Microorganisms that decompose dead or decaying matter. **134, 136, 137, 142**
Deforestation When a large number of trees are chopped down. **138, 141**
Dehydration When the water levels in the body are too low. **51, 88**
Denature When the hydrogen bonds and other bonds in a protein are broken and the protein no longer functions. **64, 136**
Dendrites The part of a neurone that receives nerve impulses from other neurones. **75**
Deoxygenated blood Blood that contains very low volumes of oxygen. **34**
Desertification When fertile land is no longer good for growing crops on. **138**
Detritivores Organisms that eat dead or decaying matter. **136**
Diabetes When either no insulin is produced, or the body stops responding to the insulin in the blood. **41, 84–5, 145**
Dialysate The liquid in the dialysis machine which helps to clean the blood of wastes. **89**
Dialysis Using a machine to clean the blood and produce urine when the kidneys are faulty. **89**
Dialysis machine A machine that cleans the blood and produces urine. **89**
Diaphragm (contraception) A barrier method worn by a woman to prevent sperm from reaching an egg during sexual intercourse. **92**
Differentiate When cells change into a specialised cell. **11, 100, 115**
Diffusion The spreading out of particles resulting in a net movement from an area of higher concentration to an area of lower concentration. **21–2**
Disease resistance Organisms that are resistant to disease. **20, 112**
Disinfectants A chemical that kills or inhibits the growth of microorganisms. **13**
Distribution The way in which a species is arranged in an area. **131, 133, 138**
DNA The molecule that genetic material is made of. **8, 18, 101–2, 104, 114**
Dominant The allele that is expressed in the phenotype. **105, 107, 108, 109, 119**
Dose The volume of a medicine that it is safe and effective to take. **56**
Double blind trial When the patient and the doctor do not know which medicine the patient is taking. **56**
Double circulatory system Circulatory system where the blood travels through the heart twice in each cycle. **34**
Double helix The coiled shape of the DNA molecule. **102**

E

Effector A gland or a muscle. **74**
Efficacy How well a medicine works. **56**
Egg The female sex cell. **98, 100**
Electrical impulse A small pulse of electricity that is carried along the neurones. **74–76, 79, 81**
Electrical stimulation Applying a weak electrical current to the brain while the patient is conscious. **78**
Embryo A developing baby in the uterus. **11, 19, 93, 98, 100**
Embryo screening When embryos are screened for genetic diseases. **109**
Emulsifies Makes large droplets of fat into smaller droplets of fat. **29**
Endocrine system The organ system that secretes hormones into the blood. **82**
Endothermic Reactions that have a net absorbance of heat energy. **63**
Energy The ability to do work e.g. move something, or carry out a reaction. **63, 68**
Engulf The ability of phagocytes to take in a pathogen inside itself. **37, 53**
Environmental change Variation that comes from the environment. **138**
Enzyme Protein that acts as a biological catalyst to speed up reactions. **29, 30–1, 33, 104, 114, 142**

Epidemiological Data which deals with the incidence, distribution, and other factors relating to health. **40**

Equilibrium When there are the same number of particles on either side of a partially permeable membrane. **21**

Eukaryotic Cells that have genetic material enclosed in a nucleus. **8, 9**

Evaporation The process of changing from a liquid into a gas. **135**

Evidence Facts that show a theory to be true. **117, 118, 120, 121**

Evolution The gradual change of simple life forms into more complex ones. **111, 117–18, 120–1**

Evolutionary tree A diagram that shows how closely related two organisms are based on their DNA sequences, physical characteristics and behaviour. **121**

Exothermic Reactions that have a net release of heat energy. **68**

Experiments A scientific procedure undertaken to make a discovery, test a hypothesis, or demonstrate a known fact. **65, 117, 119**

Extinction When no individuals of a species are alive. **121, 140**

Extremophiles Organisms that live in extreme environments e.g. volcanoes. **129**

F

Faeces The waste from digestion that is removed from the body through the anus. **28, 143**

Faulty valve When a valve in the heart no longer functions. **34**

Fermentation When sugars are broken down to alcohol by microorganisms in anaerobic conditons. **69**

Fertilisation When a sperm fertilises an egg to produce an offspring. **100**

Fertiliser Chemical or biological material to improve the growth of plants. **136**

Fever When the temperature of the human body is above 37.5°C. **50–52, 55, 58, 81**

Fish stocks The amount of fish there are in a given area. **144**

Food chain Feeding relationships within a community. **128, 130**

Food security Having enough food to feed a population. **144**

Food web Food chains overlap with other food chains to form food webs. **128, 130**

Fossil The remains or traces of prehistoric organisms. **120**

Fossil fuels Fuels that formed from fossil remains e.g. coal. **134, 138**

Fossil record The record of the evolution of living organisms as evidenced from fossils in rocks from different ages. **120**

FSH Hormone that causes the eggs in the ovaries to mature. **90, 91, 93**

Functional adaptation An adaptation that has occurred through natural selection to overcome a functional problem. **129**

Fungi A group of organisms that have eukaryotic cells with a chitin cell wall. **48, 52, 99, 134, 145**

Fungicides A chemical that kills fungi. **52**

Fuse When two cells or nuclei are joined together. **98, 100, 115**

G

Gall bladder Organ in the body that secretes bile made by the liver. **28**

Gamete A sex cell. **98, 100, 105**

Gas exchange The exchange of oxygen and carbon dioxide in the lungs. **35, 36**

Gene A short section of DNA that provides the instructions to make a protein. **101, 104–6, 113, 119**

Gene therapy When a faulty allele is repleaced with a fuctional allele. **114**

Genetic Variation that comes from your genes. **119**

Genetic counselling Advice given to people who want to have children and know that they carry a disease allele. **109**

Genetic engineering The addition of a gene from another organism into an organism's genome. **113–14**

Genetically modified organisms Organisms that have genetic material from another organism. **113, 145**

Genome All of the genes that a person has. **113**

Genotype The alleles of a gene that a pesron has in their genome. **105–7**

Geographical separation When two populations of a species are in two different areas. **118**

Geological activity The movement of rocks eg. an earthquake. **117, 120**

Geotropism When plants respond to gravity. **95**

Gibberellins A plant hormone that is used to initiate seed germination, promote flowering and increase fruit size. **95**

Global warming The gradual increase in the overall temperature of the Earth's atmosphere. **138, 140**

Glomerulus Part of the nephron where ultrafiltraton occurs. **86**

Glucagon A hormone that causes the liver to breakdown glycogen and fats into glucose. **83**

Glucose A type of sugar monomer. **63, 67–9, 83**

Glycogen An energy storage molecule found in animals. **83**

Golden rice Rice that has been genetically modificed to produce beta carotene, which is needed to make vitamin A inside the body. **113, 145**

Greenhouse effect When greenhouse gases build up in the Earth's atmosphere and absorb heat energy from the Sun. **140**

Greenhouse gases Gases which cintribute to the greenhouse effect e.g. carbon dioxide. **140, 141**

Gregor Mendel The first person to show the pattern of inheritance. **119**

Guard cells Plant cells on the underside of a leaf that open and close the stomata. **44, 45**

H

Habitats The environment in which a species normally lives. **125**

Heart attack When the blood supply to the muscles of the heart is suddenly blocked, usually by a blood clot. **38**

Heart failure Muscle weakness in the heart, or a faulty valve, means that the heart does not pump enough blood around the body at the right pressure. **38, 39**

Herbivores Animals that eat plants. **61**

Herd immunity When many people are vaccinated against a pathogen, to make it less likely that the pathogen will spread through the population. **54**

Homeostasis Keeping the internal environment at a set level, despite the external conditions. **73–97**

Hormone A protein that travels in the blood to target organs to bring about an affect. **11, 19, 37, 58, 82, 83, 88, 90–95, 115**

Hybridoma Fusing of a lymphocyte and a tumour cell. **57**

Hyperglycaemia When the concentration of glucose in the blood is too high. **84**

Hypoglycaemia When the concentration of glucose in the blood is too low. **84**

Hypothalamus Area of the brain that maintains homeostasis by sending nerve impulses to effectors. **88, 94**

Hypothermia When the body temperature is too low. **81**

I

In vitro fertilisation When the egg is fertilised by the sperm in a petri dish. **93**

Inbreeding When organisms breed and produce offspring with a close family member. **112**

Insulin A hormone that causes cells to take up glucose from the bloodstream and the liver to store glucose as glycogen. **83, 84, 145**

Interacting When factors have an effect on each other. **40–43, 64, 90, 125, 126, 141**

Interdependence When species depend on each other for survival. **126**

Interspecific competition Competition between individuals of different species. **125**

Intraspecific competition Competition between individuals of the same species. **125**

Inversely proportional When one variable increases, the other decreases in proportion. **65**

Irregular A heartbeat that not regular, too fast or too slow. **34**

K

Kidney Organ that filters the blood and removes waste to produce urine. **86–7, 88, 89**

Kinetic energy The energy of motion. **21, 30**

L

Lactic acid Substance that builds up in the muscles during anaerobic respiration and causes muscle fatigue. **68, 70**

Large intestine Organ in the body that absorbs water from food waste to make faeces. **28**

Left atrium Small chamber on the left side of the heart that receives oxygenated blood from the lungs. **34, 38**

Left ventricle Large chamber on the left side of the heart that contracts to push oxygenated blood around the body. **34, 38**

Lens A part of the eye that focuses the light onto the retina. **79, 80**

LH A hormone that controls when an egg is released from an ovary. **90, 91, 93**

Life cycle The series of changes in the life of an organism including reproduction. **52**

Lignin A waterproof molecule found in the walls of the xylem. **45, 61**

Limiting factor The factor that is limiting the rate of photosynthesis. **64**

Lower concentration When there are fewer particles in a given area. **21**

Lymphocytes White blood cells that have a specific immune response. **37, 53, 57**

M

Magnesium Mineral that is needed by plants to produce chlorophyll. **60**

Magnification How many times larger an image is compared to the specimen. **12**

Malignant tumour A growth of abnormal cells that can spread to different parts of the body and cause secondary tumours. **43**

Master gland A gland that controls all of the other glands. **82**

Medulla An area of the brain that controls involuntary movement. **78**

Meiosis Division of the genetic material to produce four genetically different daughter cells. **98, 100**

Menstrual cycle The cycle of blood lining in the uterus, ovulation, and blood shedding every 28 days (approximately). **90–1**

Meristem tissue Undifferentiated plant tissue that can become different types of specialised cells. **19, 20, 44**

Messenger RNA An RNA copy of a DNA template that contains the instructions to make a protein. **103**

Methane A greenhouse has that can be used as a fuel. **136**

Micrometres One thousandth of a millimetre. **8, 12, 16**

Microorganisms Organisms that can only be seen using a microscope. **13–14, 145**

Microvilli Tiny foldings on the surface of villi in the small intestine to increase the surface area. **22**

Migration The movement of organisms from one area to another. **138, 140**

Mineral Ions that are needed in small amounts by organisms. **60**

Mitochondria Organelle in the cell that is the site of aerobic respiration. **45, 68**

Mitosis Division of the genetic material to produce two genetically identical daughter cells. **18, 43, 98, 100**

Monoclonal antibodies Many copies of identical antibodies that are made against a specific antigen. **57–8**

Mosquito nets A narrow mesh that can be placed over beds, doors and windows to prevent mosquitoes coming in. **52**

Motor neurone Neurone that carries nerve impulses from the brain. **75**

MRI A scanning technique that gives detailed images of the body. **78**

MRSA A strain of Staphylococcus aureus that is resistant to many antibiotics. **55, 120**

Multiple births When a woman gives birth to two or more babies at a time. **93**

Muscle cells The cells that muscles are made of. **10, 68, 83**

Mutations A change in the order of bases in the DNA sequence. **43, 104, 110, 120**

Mycoprotein A type of protein made by fungi. **145**

N

Nanometres One thousandth of a micrometre. **8**

Natural selection The process whereby organisms better adapted to their environment tend to survive and produce more offspring. **110, 111, 117, 118**

Negative feedback cycle A system that increases or decreases a stimulus back to the set point. **83, 88, 94**

Nephron The part of the kidney that filters the blood and produces urine. **86, 87**

Nerve impulse An electrical impulse that travels along the neurones. **10**

Neurones The cells of the nervous system. **75, 78**

Neutralises Makes a neutral pH by adding an acid and an alkali together. **29**

Nitrates Minerals that are needed by plants to make amino acids and nucleotides.

Non-coding DNA Section of the DNA that controls the genes, switching them on and off. **104**

Non-communicable disease Disease that is not infectious. **40, 42**

Non-specific defences General defences by the body to prevent all pathogens from entering the body. **53**

Nucleotides The monomer that makes up the DNA molecule. **102**

Nucleus Organelle that contains the genetic material. **8, 18, 115**

O

Oesophagus Leads from the mouth to the stomach, also called the foodpipe. **28**

Oestrogen A female hormone that causes the development of the sex characterisitcs in women. **90, 91**

Optic nerve The nerve that leads form the back of the eye to the brain. **79**

Optimum The condition at which an enzyme works best. **30, 33, 73**

Oral contraceptives A pill containing hormones to prevent pregnancy. **92**

Organ A collection of several tissues that work together to carry out a particular function. **28–9, 34–5, 44**

Organ system A collection of several organs that work together to carry out a particular function. **10**

Osmosis The movement of water from a dilute solution to a concentrated solution through a partially permeable membrane. **23–4, 86**

Ovaries Sex organ in women that produces an egg every month for fertilisation. **90, 91**

Ovulation The time in a month when an egg is released from an ovary. **90, 91**

Oxygen debt When the lactic acid is broken down using oxygen after the exercise is finished. **68, 70**

Oxygenated blood Blood that contains high volumes of oxygen. **34**

P

Pancreas Organ that produces digestive enzymes for the small intestine and secretes insulin and glucagon into the blood. **28, 82, 83**

Parasites An organism that lives on anther organism without giving any benefit. **99, 128**

Partially permeable membrane A membrane that allows small molecules to diffuse across. **23, 89**

Pathogens A microorganism that causes disease. **37, 40, 48, 53–4, 59, 61**

Penicillin The first antibiotic that originated in a fungus called Penicillium. **55, 56**

Permeable When water or other small molecules can diffuse across a membrane. **23, 88, 89**

Peripheral nervous system The neurones that connect the limbs to the spinal cord. **74**

Pesticides A chemical that kills pests. **52**

Pests An insect or microorganism that attacks crops. **52, 59, 113, 114, 128, 144, 145**

pH A log scale of hydrogen ions which indicates if a solution is acidic, neutral or alkaline. **29, 30, 33, 137**

Phagocytosis The process of a phagocyte engulfing a pathogen. **37, 53**

Phenotype The expressed characteristics of an organism. **105–7, 110, 118**

Photosynthesis The process of converting sunlight, water and carbon dioxide into sugars and oxygen. **44, 52, 63–7, 130, 135, 138**

Phototropism When plants respond to the light. **95**

Pituitary gland The master gland in the brain that controls all of the other glands. **82, 88, 91**

Placebo A fake medicine that looks like the real medicine. **56**

Plantlet A small plant. **115**

Plasmid A small circle of DNA found in bacteria. **114**

Plasmolysed When a plant cell loses too much water and the cell membrane comes away from the cell wall. **23**

Pollution Contamination of the natural environment that can cause damage to species. **139**

Polynucleotide A chain of nucleotides. **102**

Pooter A device for trapping small flying insects. **131**

Population A group of organisms of the same species living in the same area at the same time. **125, 126, 132, 133**

Precipitation When water droplets fall as rain or snow. **135**

Predation The action of eating other organisms. **126**

Predator An organism that eats other organisms. **126**

Predator prey cycle The interaction of prey and predator species. **126**

Prey An organism that gets eaten by another organism. **126, 130**

Primary consumer Herbivores that consume the producers. **130, 142**

Probability The chances of an event happening. **106, 109**

Producer Usually plants, they convert light energy from the Sun into chemical energy, through the process of photosynthesis. **130, 142**

Progesterone A hormone that maintains the uterus lining during pregnancy. **90**

Prokaryotic Cells that do not have a nucleus. **8**

Protein A polymer made of many amino acids. **101, 103–4**

Pulmonary artery Artery that carries deoxygenated blood from the heart to the lungs. **34**

Pulmonary vein Vein that carries oxygenated blood from the lungs to the heart. **34**

Pulse rate The number of heart beats per minute. **70**

Punnett square A genetic diagram used to predict the genotypes of offspring. **107**

Pupil A small hole in the centre of the iris that allows light to enter the eye. **79**

Purified A pure sample of a product, with no contamination by another product. **57, 145**

Pyramid of biomass The relative amount of biomass at each trophic level. **143**

Q

Quadrat A 1 m by 1 m square made of wire, which may be divided into smaller squares. **131–3**

R

Radiation Energy that has travelled to Earth from the Sun. **140, 143**

Ratio The number of offspring that are expected to have a particular phenotype. **21, 22, 107**

Receptor An area of specialised cells that detect a stimulus. **74, 81**

Recessive An allele that is not expressed when a dominant allele is present. **105, 107–9, 119**

Recycling Using previously used products to make new products. **141**

Reflex arc An involuntary series of nerve impulses that do not involve the brain. **75–6**

Relay neurone A neurone that carries nerve impulses from the sensory neurones to the motor neurones in a reflex arc. **75, 76**

Renewable energy A way of producing energy that wil not run out in the future. **141**

Reproductive isolation When populations of the same species cannot reproduce with each other because of different mating behaviours. **118**

Resistant When bacteria are no longer killed by one or more antibiotics. **51, 55, 120, 121**

Resistant strain A strain of bacteria that are resistant to one or more antibiotics. **120**

Resolution/resolving power The shortest distance between two points on a specimen that can still be distinguished as separate entities. **12**

Respiration The process of using oxygen to break down glucose to produce energy, making carbon dioxide and water as byproducts. **26, 68–9, 134, 136, 143**

Restriction enzyme An enzyme that cuts the DNA every time it finds a specific sequence of bases. **114**

Retina Area at the back of the eye that is covered in light-sensitive receptors that detect light and create an image. **79, 80**

Ribosome Small organelle in the cell that translates mRNA into a polypeptide chain. **103**

Right atrium Small chamber on the right side of the heart that receives deoxygenated blood from the body. **34, 36, 37**

Right ventricle Large chamber on the right side of the heart that contracts to push deoxygenated blood to the lungs. **34, 36, 37**

S

Salivary gland A gland in the mouth that produces saliva. **28–30, 33**

Sampling Counting individuals of a species in a small area in order to estimate the population in the whole area. **40, 131, 133**

Secondary consumer Carnivores that consume the primary consumers. **130, 142**

Secondary sex characteristics Changes to the human body that occur during puberty that do not involve the sex organs. **90, 91**

Secondary tumours A tumour formed when a malignant tumour moves from its first location. **43**

Selection pressure A pressure that affects an organism's ability to survive. **117, 118**

Selective breeding When two organisms are bred together to produce offspring with desired characteristics. **112**

Selectively reabsorbed The process by which important molecules are absorbed back in the blood from the nephron. **86**

Sensory neurone A neurone that carries nerve impulses to the brain. **75**

Sex chromosomes The chromosomes that control whether an emvryo develops into a male or female. **109**

Sexual reproduction Reproduction that involves male and female gametes. **98–9**

Sieve plate The end plate of the cells in the phloem containing small pores for the sugar sap to pass through. **45**

Small intestine Organ in the body that uses enzymes to digest food and absorbs the products of digestion into the bloodstream. **22, 26, 28, 29**

Specialised cells Cells that are adapted to carry out a particular function. **10**

Specialised exchange surfaces Membranes that are adapted to allow fast diffusion. **22**

Speciation When two new species are formed from one species. **111, 118**

Species A group of similar induviduals that can breed together to priduce fertile offspring. **111, 125–6, 128, 129**

Specific The particular antigen that each lymphocytes responds to. **57, 58**

Specimen A sample of tissue. **12, 15, 117**

Spinal cord A collection of neurones that connect the peripheral nervous system and the brain. **74, 75**

Stable community When the populations within a community are well adapted to the environment. **126**

Starch An energy storage molecule found in plants. **67**

Statins Medicines that reduce blood cholesterol levels and slow down the rate of fatty material deposit. **39**

Stem cells Cells that are not differentiated and can become any type of cell. **11, 19–20**

Stent A small tube placed inside the artery that holds the artery open. **39**

Sterile An organism that cannot produce offspring, or a condition where no microrganisms are growing. **115**

Stimulus A change in the environments that can be detected by receptors. **73–75**

Stomach Organ in the body that uses acid and enzymes to digest food. **28**

Stomata Small pores on the underside of a leaf that allow gases to enter and exit the leaf. **44, 45**

Structural adaptation Adaptations to the body of the organism. **129**

Sub-cellular structures The structures and organelles within cells. **9, 11, 12, 18, 45, 98**

Sugar Monomers of polysaccharides e.g. glucose. **26, 45, 134**

Sugar-phosphate backbone Alternating sugars and phosphates on a strand of DNA or RNA. **102**

Surface to volume ratio How large an organism's surface is caompared to its volume. **21, 22**

Sustainable A method of doing something that can be maintained at a certain rate or level. **144**

Symbiosis When two species live closely together and benefit from each other. **128**

Sympatric speciation A form of speciation when the two populations are in the same area. **118**

Synapse The small gap between two neurones. **75**

T

Tertiary consumer Carnivores that eat other carnivores. **130, 142**

Testes Sex organ in men that produces sperm for fertilisation. **90**

Testosterone A male hormone that causes male sexual characteristics to develop. **90**

Thermoreceptor A receptor in the skin or hypothalamus that detects a change in temperature. **81**

Thermoregulatory centre An area in the brain that is responsible for reguating temperature in the body. **81**

Thyroid gland A gland in the body that produces thyoxine. **82, 94**

Thyroxine A hormone that increases the rate of metabolism, generating heat. **94**

Tissue A collection of similar cells that work together to carry out a particular function. **10, 25, 44, 115**

Tissue culture A sample of tissue grown on a petri dish. **95, 115**

Toxicity How toxic a medicine is. **56**

Toxins Chemicals released by pathogens which cause disease in the infected organism. **51, 61**

Trachea Leads from the throat down into the lungs, also known as the windpipe. **35**

Transect A long line made with string or a measuring tape, with quadrats placed at intervals along it. **131, 133**

Transgenic An organism that contains genetic material from another organism. **145**

Translocation The movement of sugar sap up and down the plant. **45**

Transpiration The loss of water from the top part of the plant. **45–6, 135**

Transpiration stream The movement of the water column up the xylem of the plant. **45**

Transplant When an organ is surgically implanted from another human or animal. **39, 89, 115**

Trophic levels The position it occupies in the food chain. **142–3**

Tumour A group of abnormal cells that divide uncontrollably. **43**

Turgid When a plant cell swells up with water. **9, 23, 45**

U

Ultrafiltration When high pressure in the glomerulus forces small molecules in the blood into the Bowman's capsule. **86**

Urea A less toxic byproduct of the breakdown of ammonia, that is lost from the body in urine. **87, 89**

Urine A liquid lost from the body made of urea, and excess salt and water. **71, 73, 86–88, 143**

V

Vaccination When a small amount of dead virus or antigen is introduced into a body in order to produce an immune repsonse. **51, 54**

Valves Small flaps in the heart and veins to prevent the backflow of blood. **34, 38, 39**

Variation Genetic differences between individuals of the same species. **110**

Variety The differences between individuals in a species and between species. **98**

Vector An animal that carries a pathogen, or a small amount of DNA that is used to introduce a gene into a new organism. **52, 114**

Vena cava The vein that carries deoxygenated blood to the heart and the largest vein in the body. **34**

Villi Folding of the inner membrane of the small intestine to increase the surface area. **22**

Visking osmometer Visking tubing with a glass tube in the top to measure the amount of osmosis. **23–4**

Vitamin A A vitamin that is needed by the eye to make the pigmnet needed for night vision. **113**

W

Water cycle The cycle of water from the sea, to rain, to rivers and then back to the sea. **135**

Waxy cuticle A waterproof layer on the top of a leaf. **61**

Y

Yield The volume of product that is produced. **65, 112–114, 145**

Z

Zones of inhibition An area on an agar plate where there are no microorganisms. **17**

Zygote The fertilised egg. **98, 100**

Answers

ll biology

karyotic and prokaryotic cells

In the cytoplasm as a loop of DNA and maybe as plasmids.

5 μm **3** 2 x 10^2 nm

imal and plant cells

Award one mark for each correct column:

Sub-cellular structure	Animal cells	Plant cells	Prokaryotic cells
Nucleus	✓	✓	
Mitochondria	✓	✓	
Ribosomes	✓	✓	✓
Cytoplasm	✓	✓	✓
Cell membrane	✓	✓	✓
Chloroplast		✓	
Permanent vacuole		✓	
Cellulose cell wall			✓

The more mitochondria there are, the more respiration will be carried out; Active cells need more energy.

The organism is not a plant; It has some features of plant cells/has chloroplasts/ has a cellulose cell wall; It is one-celled/ unicellular or plants are multicellular.

ll specialisation

A cell that has differentiated in order to carry out a particular function.

A nerve cell has many dendrites for passing the nerve impulse onto nearby nerve cells.

A nerve cell has a long axon for allowing the nerve impulse to travel along a part of the body.

Sperm cells are not a tissue; as the cells do not work together to carry out their function.

Xylem cells have no ends and are hollow to make a tube for water to move through; lignin in the cell wall to waterproof and give strength to the cells to stop them collapsing and water leaking out.

ll differentiation

Stem cell **2** Embryo; Plant

Cell divides; cell is exposed to a chemical/ hormone; cell changes shape/acquires new sub-cellular structures.

croscopy

Higher magnification; Higher resolution/ resolving power.

Magnification = $\frac{3\text{ cm}}{12\text{ μm}}$

Magnification = $\frac{30{,}000\text{ μm}}{12\text{ μm}}$

Magnification = x2500

Size of the image = Magnification x real size of cell

Size of the image = 12,000 × 4 μm

Size of the image = 48,000 μm or 4.8×10^4 μm

Culturing microorganisms

1 Bacteria divide by binary fission; The bacterium doubles in size and divides into two daughter cells.

2 Sterilising equipment; sterilising inoculation loop; taping lids down/storing Petri dishes upside down; culturing microorganisms at 25°C.

3 Cross-sectional area = 3.142×200^2
= $3.142 \times 40{,}000$
= 125680 μm^2 or 1.3×10^5 μm^2

Required Practical 1

1 x400 **2** 5 μm

3 Four of: Place the blood sample onto a slide; Place the slide on the stage; Make sure light is passing through the sample/light is on; Bring the blood sample into focus by looking down the eyepiece lens and moving the coarse focus; Use a higher magnification objective lens and bring the blood sample into focus using the fine focus.

Required Practical 2

1 Bacterial growth is inhibited; due to the action of an antiseptic/antibiotic.

2 No unwanted microorganisms on the agar plate which could affect results of investigation; Unwanted microorganisms could make someone ill.

3 Cross-sectional area = 3.142×0.5^2
= 3.142×0.25
= 0.7855 cm^2

Mitosis and the cell cycle

1 Growth; repair/replacement of cells; asexual reproduction.

2 At the beginning of mitosis, the chromosomes are already doubled inside the nucleus; The nucleus breaks down and the chromosomes line up in the centre of the cell; One set of chromosomes is pulled to each side of the cell to form two new nuclei; The cytoplasm and cell membranes divide to form two identical daughter cells.

3 Number of cells = 1×2^{24}
= 16777216 cells
= 1.7×10^7 cells

Stem cells

1 In the root/shoot tip.

2 Replacing cells; Development of the embryo; Medical treatment; Medical research.

3 Take cuttings from the root tip/shoot tip; Use the cuttings to produce many cloned plants; The plants would be genetically identical.

Diffusion

1 The movement of particles; from an area of high concentration to an area of low concentration.

2 Any two answers from below: Increase the surface area; Increase the temperature; Increase the difference in the concentration of the particles.

3 Surface area 4 × 4 × 6 = 96 cm^2;
Volume = 4 × 4 × 4 = 64 cm^3;
Surface are to volume ratio = 96:64 or 3:2 or 1.5:1

Osmosis

1 Water will move out of the animal cell by osmosis; The cell will shrivel and crenate.

2 Percentage increase in mass
$= \frac{(14 - 8)}{8} \times 100\%$
= 75%

3

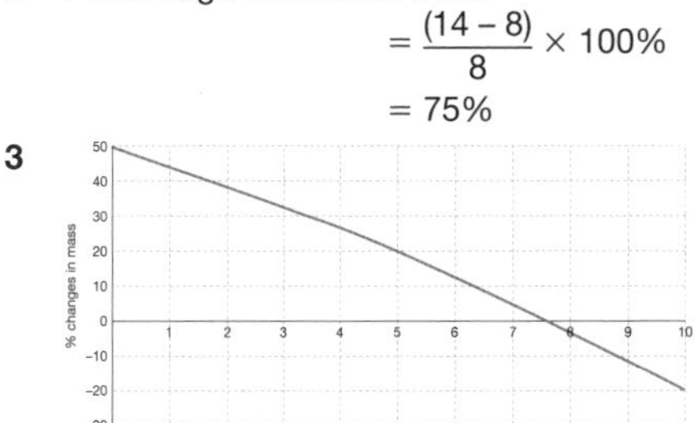

(X and Y axis drawn correctly; X axis labelled as 'Concentration of salt solution (%)' and Y axis labelled as 'Percentage change in mass); points plotted correctly; points connected together with a straight line.)

Required practical 3

1 **a** 6%

b Mass of potato cube = 5.3g

c Two from: Type of plant tissue/potato; Mass of original potato; Amount of time spent in sugar concentration;
Volume of each sugar concentration.

Active transport

1 The difference between the two concentrations; the greater the difference, the greater the concentration gradient.

2 Respiration

3 Mineral ions move from a dilute concentration in the soil, to a higher concentration in the cell; They move through carrier proteins in the cell membrane; This requires energy.

Review it!

1 No nucleus; no membrane bound organelles; has plasmids.

2 **a** Place the specimen onto the stage; Line up the objective lens and the eyepiece lens; and focus the specimen by moving the stage with the coarse focus.

b 30 000 ÷ 10; = × 3 000

3 **a** Root hair cells have long thin hairs; and do have chloroplasts; The thin hairs increase the surface area to take up water and minerals from the soil.

b Meristem tissue; in the root differentiates; into root hair cells.

4 a Use aseptic techniques; to grow bacteria on an agar plate; Make 5 wells in the agar and add 1 cm^3 of antibiotic to wells B – E. Add nothing to well A; After two days, observe the radius of the colonies.

b Independent variable is the type of antibiotic; Dependent variable is the radius of each colony.

c A 10.2; B 0.1

d B was the most effective antibiotic; because the cross-sectional area of the colony treated with B is the smallest.

e To prevent other microorganisms growing on the agar plate; To prevent the student becoming ill.

f Carry out the investigation at least three times; Make sure the same volume of antibiotic is used each time; Make sure all of the agar plates are kept at the same temperature / are left for the same length of time.

5 a Prophase; Metaphase; Anaphase; Telophase.

b Anaphase

c Because onion root tips are growing; Root tips contain meristem tissue.

6 a Undifferentiated cells; that can become any type of specialised cell.

b Embryos / umbilical cords; Adult organs contain a few stem cells.

c No rejection of cells / organs by patient; Transfer of viral infection; No waiting time for transplants; Ethical / religious objections.

7 a Diffusion is the movement of particles from an area of high concentration; to an area of low concentration.

b The salt ions will move towards B.

c The salt ions are more concentrated on the A side of the membrane; and less concentrated on the B side of the membrane.

8 a Facilitated diffusion.

b In diffusion, particles move from an area of high concentration to an area of low concentration, but in active transport, the particles move from an area of low concentration to an area of high concentration; Diffusion is a passive process, but active transport requires energy.

Tissues, organs and organ systems

The human digestive system

1 In the pancreas.

2 To break down starch/carbohydrate; into sugars.

3 Bile emulsifies lipids into smaller droplets; This gives a larger surface area; for the lipase enzymes to digest the lipids.

Enzymes

1 pH3 is lower than the optimum pH for amylase, which is pH7; Amylase would lose activity/rate of reaction would decrease.

2 20/4 = 5 cm^3/min

3 The hydrogen bonds in the enzyme are broken; The active site is no longer the correct shape/complementary to the substrate; The substrate would no longer be able to bind to the enzyme's active site.

Required practical 4

1 Contained protein.

2 Add ethanol to a food sample; Add distilled water to the food sample and observe if an emulsion formed.

3 Iodine would change pasta a blue-black colour; Benedict's reagent would remain blue; Pasta contains starch, but not sugars.

Required practical 5

1 Breaks down/digests starch (amylose) into sugars (maltose).

2 The rate of reaction would decrease; because the amylase loses activity above the optimum pH.

3 Rate of reaction = 0.033 cm^3/s

The heart

1 Pulmonary vein

2 To prevent the blood flowing backwards.

3 A group of cells in the right atrium maintain the resting heart rate; they make sure that the ventricles contract shortly after the atria; it is important to prevent the heart rate from becoming irregular.

The lungs

1 In the blood capillary.

2 The gases have less distance to travel; This speeds up the rate of diffusion.

3 The blood capillaries move the oxygen away from the alveoli; This means that the concentration of oxygen in the capillary next to the alveoli will always be lower than the concentration of oxygen inside the alveoli; This means that there is a greater concentration gradient between the alveoli and the blood capillaries.

Blood vessels

1 Vein 2 Aorta

3 Capillary walls are one cell thick; to allow a short diffusion pathway for gas exchange; in the tissues and around the alveoli.

Blood

1 Any two from: plasma; red blood cells; white blood cells; platelets.

2 Biconcave in shape; no nucleus; packed with haemoglobin.

3 The endoplasmic reticulum is the site of protein synthesis in the cell; Lymphocytes need to make many antibodies, which are a type of protein.

Coronary heart disease

1 Any two from: angina; heart attack; heart failure.

2 Build-up of fatty deposits and cholesterol; in the coronary arteries.

3 Blood would flow backwards; back into the heart/into the previous heart chamb[er]; The heart would have to pump harder t[o] pump blood through the heart.

Health issues

1 Positive correlation

2 As a histogram; because it is showing numerical data.

3 Fewer white blood cells; means that the person is less able to fight off pathogen[s].

Effect of lifestyle on health

1 Any two from: diet; smoking; exercise.

2 Any one from: cost of medicines; cost o[f] hospital/doctor visits; time off work/not able to work.

3 Percentage decrease = 23%

Cancer

1 Any two from: genetic; smoking; diet; alcohol; ionising radiation.

2 Abnormal cell divides without being checked; Cells divide out of control to form a large mass.

3 More tumours to remove; More difficult to find the tumours in the body; Tumours may be impossible to remove by surgery; Medicine/chemotherapy/radiotherapy is needed.

Plant tissues

1 Xylem

2 4–5 cells drawn correctly; cells correctly labelled with cell wall, cell membrane, permanent vacuole, chloroplast, cytoplasm and nucleus; label lines draw[n] with a straight line.

3 To allow light to get to the palisade mesophyll/tissue; for photosynthesis.

Transpiration and translocation

1 Water moves into the root hair cells by osmosis; Water moves up the xylem; Water diffuses out of the leaves through the stomata.

2 Increasing temperature; Increasing light levels; Increasing air movement; Decreasing humidity.

3 Rate of transpiration $= \frac{\text{water lost}}{\text{time}}$
$= 3$ cm^3/hour

Review it!

1 a Any two from: Amylase / carbohydras[e]; pepsin / protease / trypsin; lipase

b Sugars

c The enzyme's active site is complementary to the shape of the substrate; The enzyme binds to the substrate; to make a product.

d The enzyme would become denature[d]

2 a Add Benedict's reagent; and place into a hot water bath for a few minutes; If sugar is present, the solution will change to green / yellow / orange / brick-red.

b Add iodine solution to the solution; I[f] starch is present, the solution will tu[rn] blue-black.

c Add Buiret's reagent to the solution; If protein is present, then the solution will turn lilac.

a i Vena cava ii Pulmonary vein

b The atria contract to move the blood into the ventricles; The valves in the heart prevent the backflow of blood.

c A small area of specialised cells in the right atrium acts as a pacemaker.

a Down the trachea; down the bronchi; then the bronchioles; and then into the alveoli.

b The gaseous exchange surface is one-celled thick to allow for a short diffusion pathway. There are many alveoli to give a large surface area; The blood carries the oxygen away from the alveoli to create a steep concentration gradient for oxygen.

c 10 ÷ 0.4 = 25 mm / s

a The coronary artery is blocked; and oxygen cannot reach the muscle of the heart.

b Stents hold the coronary arteries open, and allow blood flow to the heart, and statins reduce blood cholesterol levels by slowing down the rate of fatty material deposit; However, neither of these is a permanent solution. The fatty material will eventually build up if the patient does not eat a healthy diet, and the artery will need to be bypassed.

a To allow oxygen and carbon dioxide to diffuse in and out of the leaf; To allow water vapour to diffuse out of the leaf.

b i

emperature	Number of open stomata			Mean number of open stomata
	1	2	3	
0	14	16	12	14
0	34	36	38	36
0	49	51	44	48
0	87	78	81	82

ii Add °C to column header of first header.

iii The higher the temperature, the higher the number of open stomata.

iv The rate of photosynthesis increases with increased temperature; More carbon dioxide will be needed for photosynthesis, so the stomata are open to allow the carbon dioxide in.

c Increased temperature; increased light; and increased wind speed.

ection and response

mmunicable diseases

Any two from: Common cold; Influenza; HIV; Measles; Any other virus.

Direct contact with person/animal/sharp object; Through air/coughing/sneezing/ water droplets; In food/water.

3 Close contact with other people makes it easier to spread viruses/bacteria; Fungi like damp conditions so more likely to have a fungal disease; Insects like damp conditions and may carry diseases.

Viral diseases

1 In water droplets/coughs/sneezes; Direct contact with infected person.

2 Flu-like virus in the beginning; HIV attacks the cells of the immune system; Immune system stops working/unable to function.

3 TMV spreads by direct contact with plants; Removing the infected plants will reduce the spread of TMV; Uninfected plants may still be susceptible to TMV in the soil.

Bacterial diseases

1 Antibiotics

2 Salmonella can be found in poultry; To prevent the spread of Salmonella.

3 The bacteria that causes gonorrhoea could be resistant to all antibiotics; It would not be possible to treat gonorrhoea; There would be more emphasis on preventing gonorrhoea infection.

Fungal and protist diseases

1 Antimalarial drugs

2 Prevents mosquitos from biting; Mosquitos carry the protist that causes malaria.

3 Remove any infected leaves; Spray the roses with fungicide.

Human defence systems

1 Any two from: Skin; Mucus in the nose, trachea and bronchi; Cilia in the trachea and bronchi; Hydrochloric acid in the stomach.

2 Phagocyte engulfs the pathogen; The cell membrane extends around the pathogen; Enzymes inside the phagocyte digest the pathogen.

3 Lymphocytes produce antibodies; against the toxins produced by bacteria; These antibodies are called antitoxins.

Vaccination

1 Dead/inactive pathogen

2 When most of a population is vaccinated against a pathogen; non-vaccinated people are protected from infection with that pathogen.

3 TB is not common in the UK; due to vaccination programmes in the past; TB vaccinations are given only if there is an outbreak of TB.

Antibiotics and pain killers

1 Penicillin; *Any antibiotic*.

2 Antibiotics only kill bacteria; Antibiotics would have no affect against/cannot kill viruses.

3 Percentage increase = 350%

New drugs

1 Found in plants/microorganisms; Make new compounds from already existing ones.

2 Drugs are tested on cells; Drugs are tested to check their toxicity/efficacy/dosage.

3 Need to make sure that the new drug works; better than the current drug.

Monoclonal antibodies

H1 Lymphocytes

H2 The shape of the monoclonal antibody is complementary; to the shape of the antigen.

H3 The mouse antibodies would act as an antigen; The rabbit lymphocytes would be stimulated; into making monoclonal antibodies that were complementary to the mouse antibodies.

Monoclonal antibody uses

H1 Any two from: Diagnosis; Blood testing; Pathogen detection; Researching a specific protein; Treating diseases.

H2 Monoclonal antibodies are specific to cancer cells; Healthy cells will not be affected/harmed.

H3 Make monoclonal antibodies against an antigen/binding site on a protein; Add monoclonal antibodies to blood sample; If antibodies bind to antigen/binding site, then protein is present.

Plant diseases

1 Fungus

2 Nitrates; Because nitrates are needed to make proteins for growth.

3 As the years go on, the number of plants infected with TMV increases; because the TMV virus is spreading from the infected plants; The TMV virus is also present in the soil; The farmer should remove all of the infected plants; Plant TMV-resistant strains of tobacco plants.

Plant defences

1 Antibacterial chemicals; Poisons/toxins

2 The plant imitates/looks like something; that the herbivore doesn't want to eat.

Review it!

1 a A microorganism that causes disease.

b Any two from: Tuberculosis; tetanus; cholera; Salmonella food poisoning; gonorrhea.

c Any two from: Airborne / coughing and sneezing; direct contact with infected person / animal / sharp object; in food and water.

d Antibiotics

2 a Bacteria are much larger than viruses; bacteria are living cells, whereas viruses are not living / envelopes surrounding some genetic material.

b i Red skin rash; fever.

ii Vaccination; isolating infected people from non-infected people.

c i Tobacco mosaic virus.

ii TMV causes large black patterns to appear on the leaves; that prevents the leaves from photosynthesizing; This means that the plant produces less sugars and has less energy for growth.

3 a Protist / plasmodium.

b i The number of deaths from malaria decreases worldwide from 2000 to 2015; The number of deaths from malaria is highest in Africa.

ii Increased use of mosquito nets / pesticides / antimalarial drugs; The mosquito that carries the protist that causes malaria survives very well in the climate in Africa.

4 a Any two from: Skin is a physical barrier against pathogens; breaks in the skin form scabs; sweat glands in the skin produce sweat that inhibits pathogens; small hairs and mucus in the nose trap airborne particles; mucus in the trachea traps pathogens and is moved up to the throat by small hairs called cilia; hydrochloric acid and protease in the stomach kills pathogens; phagocytes kill any pathogen that invades the body.

b Phagocytes are non-specific and lymphocytes are specific; Phagocytes engulf pathogens by phagocytosis and digest the pathogen; Lymphocytes produce antibodies that attach to pathogens and prevent them from entering cells / target the pathogen for phagocytosis / act as an antitoxin; Lymphocytes have memory cells that remember the specific pathogen and respond more quickly if the same pathogen enters the body again.

5 a i A and B kill approximately the same number of bacteria; C kills more than three times the number of bacteria that A and B kills.

ii The bacteria have become resistant to antibiotics A and B; Antibiotic C is a stronger antibiotic than A and B.

b Bacteria eventually become resistant to antibiotics; The antibiotics become less effective at treating infections.

Bioenergetics

Photosynthesis

1 Glucose; Oxygen

2 In the palisade mesophyll cells; because they have the most chloroplasts; to absorb a greater amount of sunlight.

3 The six carbon atoms are from the six molecules of carbon dioxide; The 12 hydrogen atoms are from the six molecules of water; The six oxygen atoms are from the six molecules of water.

Rate of photosynthesis

1 The factor that limits the rate of photosynthesis.

2 The rate of photosynthesis would decrease; Carbon dioxide is needed for photosynthesis to occur.

3 Rate of photosynthesis = 7 cm^3/min

Required practical 6

1 At 0 and 10 cm from the plant, the rate of photosynthesis remained constant; After 10 cm from the plant, as the distance from the lamp increased, the rate of photosynthesis decreased.

2 Any two from: Temperature; The species/size of plant; The type/brightness of lamp; The time.

3 Any three from: Repeat the investigation; Use light probes to accurately measure the light intensity; Use oxygen probes to accurately measure the volume of oxygen released; Carry out the investigation in a dark room with the lamp as the only light source.

Uses of glucose

1 In the leaves.

2 Eating plants/fruit/vegetables.

3 More glucose is made/eaten than can be used at one time; Storage allows glucose to be used at a later time; Starch and glycogen can be broken down into glucose by enzymes.

Respiration

1 In the cytoplasm and the mitochondria.

2 The yeast carries out anaerobic respiration; Anaerobic respiration produces ethanol and carbon dioxide; The ethanol is the alcoholic content of wine, and carbon dioxide bubbles makes the bread rise.

3 There is a positive correlation between the temperature and the amount of carbon dioxide released; As the temperature increases, the volume of carbon dioxide produced increases.

This is because the rate of respiration in yeast increases with temperature, up until the optimum temperature.

Response to exercise

1 12 x 4 = 48 beats per minute

2 To take in more oxygen; To supply the blood with oxygen more quickly.

3 The sprinters have been respiring anaerobically; They have a build-up of lactic acid in the muscle cells; The oxygen is needed to break down the lactic acid into glucose.

Metabolism

1 Any one from: Any example where a complex molecule is made from simpler molecules; Any example where a complex molecule is broken down into simpler molecules.

2 For growth; Repair or replacement of cells or tissue; For energy storage as carbohydrates or lipids.

3 In respiration, glucose is broken down; Carbon dioxide and water are made.

Review it!

1 a $C_6H_{12}O_6$

b Any two from: Used in respiration; converted into glycogen for storage; used to produce fat for storage; combined with nitrogen to make amino acids.

c Animals store glucose as glycogen, whereas plants store glycogen as starch. Animals use glucose to produce fat for storage whereas pla use glucose to make oils for storage

2 a The breathing rate increases to oxygenate the blood more quickly. The breath volume increases to tak in more oxygen with each breath. T heart rate increases so that the bloo flows to the cells more quickly.

b Measure the pulse at the wrist or neck for 15 seconds. Multiply this number by 4 to get the pulse rate p minute.

c After anaerobic exercise, there is a build up of lactic acid in the muscle The lactic acid is transported to the liver where it is converted into gluc using oxygen.

3 a Any example that describes the formation of a complex molecule from one or more simpler ones, or t breaking of a complex molecule int simpler ones.

b i glucose + oxygen → carbon dioxide + water

ii Unbalanced: $C_6H_{12}O_6 + O_2 \rightarrow CO_2$ H_2O

Balanced: $C_6H_{12}O_6 + 6O_2 \rightarrow 6CO_2$ $6H_2O$

4 a Glucose and Oxygen

b Any two from: Increasing light intens increasing temperature increasing concentration of carbon dioxide

c When the rate of photosynthesis remains constant as one of the fact that affects the rate of photosynthe is limiting photosynthesis.

5 a Award two marks for shape of grap as below.

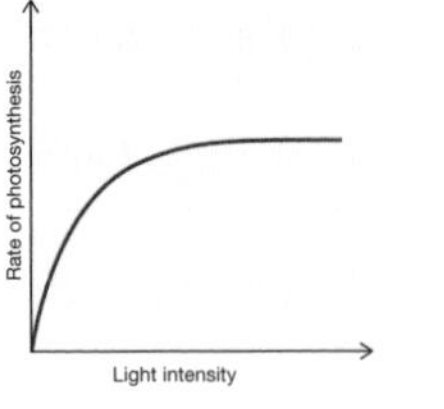

ii X axis labelled as light intensity.

Y axis labelled as rate of photosynthesis.

b As the light intensity increases, the rate of photosynthesis increases, u the rate of photosynthesis remains constant at increasing light levels because the rate of photosynthesis is being limited by a limiting factor / temperature / carbon dioxide concentration.

6 a From top to bottom of rate of photosynthesis column: 40, 30, 10,

b 30°C

c i As the temperature increase, the is more kinetic energy and the enzymes and substrates involve in photosynthesis are more likely collide.

ii After 30°C, the enzymes involved in photosynthesis start to denature. The active site of the enzyme can no longer bind to the substrate.

d Repeat the investigation at least two more times.

meostasis and response

meostasis

The maintenance of a constant internal environment.

Any one from: temperature; blood glucose levels; water levels.

The coordination centre receives nerve impulses; from the receptor; the coordination centre then sends nerve impulses to the effector; to bring about a response which will return the condition to normal.

e human nervous system

Nerve cells/neurones

To receive electrical impulses from the receptors; To send electrical impulses to the effectors.

The central nervous system is made up of the brain and the spinal cord; The peripheral nervous system is made up of all the other nerve cells; The role of the central nervous system is to coordinate the electrical impulses; The role of the peripheral nervous system is to send and receive electrical impulses.

flexes

An automatic response.

Any two from: Blinking; Pupil contracting; Coughing; Knee jerk; *Any other reflex.*

The receptors in the eye send an electrical impulse along the sensory neurone; to a relay neurone in the spinal cord; The relay neurone passes the electrical impulse to a motor neurone; which sends an electrical impulse to the muscles in the eye to make the pupil contract.

quired practical 7

The reaction time

To make the investigation reliable/To remove any anomalies from the data.

Caffeine stimulates the nervous system.

e brain

Cerebral cortex.

Coordination of movement; balance

Can only use case studies of people with brain accidents; Have to be careful not to damage brains; Brains are difficult to visualise.

e eye

The ciliary muscles.

Protects the eye; Keeps the round shape of the eye

Rods and cones both detect light; Cones detect colour, but rods do not; Cones work best in bright light, but rods work best in low light.

Focusing the light

1 Hyperopia is when near objects are not focused clearly; Corrected with a convex lens to focus the image behind the retina.

2 When the eye is looking at a near object, the ciliary muscles contract; This causes the suspensory ligaments to loosen and the lens becomes thicker; When the eye is looking at a distant object, the ciliary muscles relax; The suspensory ligaments tighten, and the lens becomes thinner.

Control of body temperature

1 They have a fever/hyperthermia.

2 The sweat evaporates from the surface of the skin; and takes heat energy with it into the environment.

3 Thermoreceptors detect a change in temperature; The thermoreceptors in the skin then send electrical impulses to the thermoregulatory centre in the brain; The thermoreceptors in the thermoregulatory centre monitor the blood directly.

Human endocrine system

1 The adrenal gland.

2 One mark for the endocrine gland, one mark for the correct hormone.

3 Both systems send messages to target organs or cells; The endocrine systems acts more slowly than the nervous system.

Control of blood glucose concentration

1 The pancreas

2 Excess glucose is stored as glycogen; in the liver and muscle cells; When the blood glucose concentration is low, glucagon causes glycogen to be broken down into glucose.

H3 When the blood glucose level is too high, insulin is released; The cells take up glucose and the blood glucose levels decrease to the normal level; When the blood glucose level is too low, glucagon is released; Glycogen is broken down into glucose and the blood glucose levels increase to the normal level.

Diabetes

1 Insulin injections

2 Any two from: shaking; dizziness; coma.

3 Any two from: eat a healthy; balanced diet; do plenty of exercise; maintain a healthy weight.

Maintaining water and nitrogen balance in the body

1 Any two from: water; ions; urea; glucose.

2 Water will diffuse out of the cells by osmosis; The cell will shrivel/crenate; The cell would no longer be able to function.

3 More water would be lost in the urine; The urine would be less concentrated.

ADH

H1 To make the collecting duct in the kidneys; more permeable to water.

H2 The collecting duct would be more permeable to water; More water would be reabsorbed back into the blood; The urine would be more concentrated.

H3 The water level is maintained at a set point; If the water level in the blood decreases below the set point than ADH is released; The water level returns to the set point as more water is reabsorbed in the kidneys.

Dialysis

1 To remove excess ions and urea from the blood.

2 Any two from: No need for surgery; Available/no waiting list; No need to take immunosuppressant drugs.

3 The blood has a higher concentration of urea than the dialysate; The urea moves from the blood into the dialysate by diffusion; down a concentration gradient.

Hormones in reproduction

1 Testosterone

2 To cause the release of a mature ovum.

3 The level of progesterone would remain high; The level of oestrogen would remain high; The level of FSH would be inhibited; The level of LH would be inhibited.

Contraception

1 Any one from: Condoms/diaphragms; Spermicidal agent; Intrauterine device; Surgical method; Abstinence.

2 Oral contraceptives contain oestrogen and progesterone; which inhibit FSH and prevents the eggs maturing.

3 Any two from: Lasts for a longer time; Don't need to remember to take it every day; Fewer side effects.

Using hormones to treat infertility

H1 FSH and LH

H2 Eggs are removed from the mother's ovaries; Eggs are fertilised by the father's sperm in the laboratory; Embryos are implanted into the mother's uterus.

H3 Multiple births are a health risk to mother and babies; IVF is emotionally and physically stressful; Not all embryos are implanted/some embryos are discarded.

Negative feedback

H1 The endocrine system

H2 Any two from: increase in heart rate; increase in breathing rate; blood diverted from skin and intestines to muscles; glycogen broken down into glucose; pupils dilated.

H3 Adrenaline is not produced by the body all of the time; and so does not have a set level; Adrenaline is only produced at times of fear/stress.

Plant hormones

1 When a plant shoot tip bends/grows towards the light.

2 Ethene

3 In phototropism, auxin gathers on the shaded side of the shoot tip; In geotropism, auxin gathers on the lower side of the shoot tip; In both geotropism and phototropism, the auxin causes the cells in the shoot tip to elongate.

Required practical 8

1 Phototropism

2 Auxin causes roots to grow downwards and causes shoots to grow upwards and towards the light.

3 The leaves in the shoot contain chloroplasts/chlorophyll; that absorb sunlight for photosynthesis; Growing towards the light means that more sunlight can be absorbed.

Review it!

1 a Homeostasis keeps the internal conditions of the body the same; whatever the outside environment may be.

b i Thermoreceptors

ii Hypothalamus

iii Sweat glands / hair erector muscles / smooth muscle in the blood vessels / skeletal muscles for shivering.

c Increased blood glucose concentration → pancreas releases insulin → cells take up insulin and liver converts glucose to glycogen → blood glucose concentration returns to normal.

Decreased blood glucose concentration → pancreas releases glucagon → liver converts glycogen to glucose → blood glucose concentration returns to normal.

d Cells and liver do not take up glucose; so blood glucose levels remain high.

2 a i Adrenal gland ii Ovary

iii Pancreas

b When the volume of water in the blood is low, ADH is released from the pituitary gland; ADH increases the permeability of the collecting ducts in the kidney so that more water is reabsorbed and less urine is produced; When the volume of water in the blood is high, less ADH is released from the pituitary gland; The permeability of the collecting ducts in the kidney is decreased so that less water is reabsorbed and more urine is produced.

c i The blood moves through the dialysis machine, where it is separated from the dialysate by a partially permeable membrane; The concentration of urea is higher in the blood compared to the dialysate, so this moves into the dialysis fluid down a concentration gradient by diffusion; If the ion concentration in the blood is higher than normal, any excess ions also move into the dialysate.

ii If a person had a kidney transplant then they would not have to have hours of dialysis every week; However, there is a long wait for a kidney transplant/risk of rejection/ need to take immunosuppressant drugs.

3 a i Cerebral cortex

ii Medulla

b i Diagram to show light focused in front of the retina; Diagram to show image upside down after the lens.

ii Concave lens

4 a Any two from: Sensory neurone; motor neurone; relay neurone.

b A voluntary nerve response goes to the brain whereas a reflex does not; Reflexes are faster than voluntary nerve responses; Reflexes involve the relay neurone whereas voluntary nerve impulses do not.

c i No; because there is not enough data to come to this conclusion.

ii Any two from: They should repeat their investigation; They should use a range of volumes of caffeine; They should test more people.

Inheritance, variation and evolution

Sexual and asexual reproduction

1 Any two from: Binary fission; Budding; Runners/bulbs/tubers.

2 The first cell of a new organism; when the nuclei of two gametes fuse.

3 Only one parent needed; More time/ energy efficient; Many identical offspring can be produced quickly.

Meiosis

1 Six chromosomes.

2 A type of cell division that produces two genetically identical daughter cells.

3 Each gamete contains half of the chromosomes as the body cell; The nuclei from a male and female gamete fuse; The fused nuclei contain the full number of chromosomes.

DNA and the genome

1 All of the genetic material of an organism.

2 A DNA molecule has two strands; that are wound around each other to form a double helix.

3 Find out about genes that cause disease; and how to treat inherited disorders; Look at genes to find out about human migration patterns from the past.

DNA structure

1 A sugar, phosphate group and a base; The phosphate group and the base are bound to the sugar.

2 The bases are joined together by hydrogen bonds.

3 A set of three bases codes for an amino acid; The sequence of the bases gives the order of the amino acids needed to make a particular protein; Each protein has a different sequence of amino acids, and so a different sequence of bases.

Protein synthesis

H1 To make a copy of the DNA code; To carry the code out of the nucleus to the ribosomes.

H2 Proteins have different shapes, depending on their function; Different amino acid sequences fold in different ways.

H3 The three base code would alter; Differe amino acids would be added from the point of the mutation; The protein may r function.

Genetic inheritance

1 There are two dominant alleles.

2 a X^cX; XX; X^cY; XY.

b One out of four children could be a b with colour blindness; 0.25 probabilit

Punnett squares

1 An allele that is always expressed.

2 3:1

H3 All of the children will have the genotype, C

Inherited disorders

1 Polydactyl

2 XY

3 25%

Variation

1 Any one from: Height; Weight; Amount c pigment in skin; *Any reasonable answer.*

2 A change in the base sequence in DNA.

3 The individual with the advantageous mutation is more likely to survive in the environment; The individual is more likely to have offspring; The mutation wi spread through the population.

Evolution

1 A group of similar individuals that can breed together; and produce fertile offspring.

2 These individuals would be less likely to survive; and less likely to have offspring

3 The two populations would not interbree Each population would undergo natural selection; The phenotypes of each population would become very different The two populations would no longer be able to interbreed with each other to produce fertile offspring.

Selective breeding

1 Any one from: Increased yield in milk; Increased size of cauliflower heads; Increased muscle mass on animals; Disease resistance in plants; *Any sensib answer.*

2 Can eliminate disease; Increase yield.

3 Breeding is tightly controlled; Only organisms with the best characteristics are allowed to breed.

Genetic engineering

1 Any one from: Disease-resistant plants; Plants that produce bigger fruits; Bacter that produce human insulin; Gene thera to overcome some inherited disorders; *Any sensible answer*.

2 A gene from another organism is cut out of its genome; The gene is placed inside a plasmid/vector; The plasmid/vector is placed inside the cells of the organism.

3 The normal allele is inserted into the genon of the person with the disorder; This overcomes the effect of the faulty allele.

ning

Any one from: cutting; tissue culture.

To prevent the growth of fungi/bacteria on the agar plates.

In a changed environment, the individuals that have characteristics that are more suited to the environment are more likely to survive; There is no genetic variation within the population; If the clones are not suited to the environment, none of them will survive.

eory of evolution

Lamarck's theory of evolution.

Something that causes an organism to struggle to survive.

It challenged the idea that God made all of the animals and the plants that live on earth.

There was not enough evidence at the time to support the theory; The mechanism of the inheritance of genes was not known for 50 years after the theory was published.

eciation

Geographical separation; reproductive separation.

Proposed theory of evolution by natural selection independently of Darwin; gathered evidence for evolution worldwide; published joint papers with Darwin on evolution.

e understanding of genetics

The pattern of inheritance in pea plants; Unchanged 'units' are passed from parents to offspring.

The dominant allele.

The structure of DNA was not known; Genes had not been discovered; The behaviour of chromosomes had not yet been observed.

dence for evolution

The remains of an organism from millions of years ago found in rocks.

Any two from: Loss of habitat; Loss of food sources; Hunting of the species.

The bacteria that are resistant to the antibiotic survive; and are more likely to have offspring; The offspring also have the resistance gene.

ssification

Phylum; Order; Family.

The two-name system of naming organisms; The genus and the species.

The archaea are more closely related to eukaryotes than bacteria; Archaea are different biochemically to bacteria.

view it!

a Reproduction without another parent/ no fusion of gametes; using mitosis.

b There are two daughter cells made by mitosis, and four by meiosis; The daughter cells made by mitosis are genetically identical, whereas the daughter cells made by meiosis are genetically different; Daughter cells made by meiosis have half of the genetic information of daughter cells made by mitosis.

c A male and a female gamete join together and the two nuclei fuse; When this happens, the new cell / zygote has a full set of chromosomes; The zygote divides by a type of cell division called mitosis.

2 a A short section of DNA that codes for a particular protein.

b A large structure in the nucleus made of many genes.

C DNA is a polymer made from four different nucleotides; Each nucleotide consists of a sugar and phosphate group with one of four different bases; The long strands of DNA consist of alternating sugar and phosphate sections; joined by the base pairs in a double helix.

d i 26% ii 24%

e TACGGCAATTAG

3 a Genetic; Environmental.

b There is variation within a species. Individuals with characteristics that are more suitable for the environment are more likely to survive; and have offspring with these same characteristics.

4 a Each species has a two part name; describing their genus and species.

b Kingdom – Animal; Genus – Canis.

5 a i

	B	b
B	BB	Bb
b	Bb	bb

ii 1:2:1

b 50% probability.

H 6 a The introduction of a gene from one organism; into another organism.

b The desired gene is cut out the original genome using enzymes called restriction enzymes; The same enzymes are used to cut open a plasmid; The gene is inserted into the plasmid; The plasmid is inserted into the bacterial cells.

c Could produce medicines more quickly / cheaply; Could insert a functional gene into a cell with a faulty gene.

Ecology

Communities

1 A group of populations living and interacting with each other in the same area.

2 Interspecific

3 The population of frogs would increase, as they would no longer be eaten.

The population of mayfly nymphs would decrease as there would be more frogs to eat them.

Abiotic factors

1 Any two from: Temperature; Light intensity; Moisture levels in the soil/air; pH of the soil; Wind intensity and direction; Carbon dioxide levels; Oxygen levels of the water.

2 Some plant species cannot grow in high or low pH; The number of plant species may be decreased.

3 The wind will remove water from the leaves of the tree; The plants may not receive as much water for photosynthesis as they would on a non-windy hillside; The plants' growth may be decreased.

Biotic factors

1 Any two from: food availability; predators; microorganisms in the soil; pathogens; competition within and between species; parasites; symbiosis; pollination.

2 Loss of habitat; Loss of food sources; The populations of other species would decrease in number.

3 The number of pea plants would decrease due the decrease in pollination by bees; The number of caterpillars would decrease due to the decrease in pea plants as a food source; The number of sparrows would decrease due to the decrease in caterpillars as a food source.

Adaptations

1 Any one from: thick fur; white fur; sharp claws; thick layer of blubber; *Any sensible answer.*

2 A behavioural adaptation is a change in behaviour to aid survival in the environment; A functional adaption is an adaptation that has evolved to overcome a particular problem.

3 They are adapted to living at high temperatures; They convert inorganic compounds into energy.

Food chains

1 An organism that feeds on primary consumers.

2 Grass → rabbit → fox (arrow pointing in the correct direction; correct organisms)

3 The energy decreases as it goes through each level on the food chain; There would not be enough energy to support another organism in the food chain.

Measuring species

1 Place the quadrat down randomly in an area; Record the types of species and the number of each species in the quadrat.

2. $25 \div 5 = 5$; $5 \times 100 = 500$ buttercups

3 So that the animals will not be easier to spot by predators; So that the animals will be able to move about in a natural way.

Required practical 9

1 Use the animal traps to capture animals in a given area; Mark the captured animals and release them; Wait a period of time and then capture animals in the same area and count how many are marked; Use the number of captured and marked animals to estimate the population size.

2 16.5 snails

3 Deaths, or migration out of the area will decrease the population size; Births, or migration into the area will increase the population.

The carbon cycle

1 Any one of: aerobic respiration; combustion.

2 Microorganisms decompose decaying matter; Microorganisms respire carbon dioxide back into the atmosphere.

3 Increased amount of combustion of fossil fuels; Decreased amount of photosynthesis; due to increased deforestation.

The water cycle

1 Evaporation; Transpiration.

2 Heat energy from the Sun warms the water droplets on the surface of the ocean; Heating changes the state of the water from a liquid to a gas which evaporates into the atmosphere.

3 As the wind transports clouds over mountains, the clouds rise; The water in the cloud cools and precipitates on the land on one side of the mountain; The cloud contains little rain by the time it reaches the land on the other side of the mountain.

Decomposition

1 Any one from: Maintain an optimum temperature; Provide a source of water; Make sure the decaying matter gets plenty of oxygen; Introduce detritivores to the decaying matter.

2 Rate of decay = 80g/day

3 Keep the tomatoes in cold conditions; Keep the tomatoes dry; Remove all oxygen from the air.

Required practical 10

1 As the temperature increases from 20°C to 40°C, the rate of decay increases; As the temperature increases from 40°C to 80°C, the rate of decay decreases.

2 The time taken for the pH to change.

3 The proteins and fats in the milk are broken down by microorganisms/ decomposers; into amino acids and fatty acids and glycerol; This decreases the pH of the milk/makes the milk more acidic.

Impact of environmental change

1 Any one of: water availability; temperature; atmospheric gases.

2 As the temperature rises, the water evaporates from the land; There is not enough water to support living organisms, and the land becomes a desert.

3 Some plant species would not be able to survive; so there would be fewer food sources for the animals; There may be less water available for the animals; The animals would need to migrate to a habitat where there was more food and water.

Biodiversity

1 Air pollution

2 Trees and plants are removed to make space to build houses; This removes habitats and food sources for other organisms; Animal species will migrate away from the area to find food and shelter.

3 There is a large amount of carbon in peat bogs; When peat is removed and burned, it releases carbon dioxide into the atmosphere; The carbon dioxide in the atmosphere is taken up by plants for photosynthesis.

Global warming

1 Carbon dioxide; Methane.

2 Radiation from the Sun warms the Earth; The heat is reflected from Earth where is absorbed by the greenhouse gases in the atmosphere; The temperature of the atmosphere increases.

3 Decreasing the use of fossil fuels in vehicles; Decreasing the use of fossil fuel power stations; Increased use of renewable energy; Decreased deforestation.

Maintaining biodiversity

1 Any one of: Combustion of fossil fuels; Pollution; Deforestation.

2 Hedgerows provide habitats and food sources for species; so the number of species in the ecosystem increases.

3 Limit numbers of tourists allowed each year; Recycle waste; Reduce pollution from tourists.

Trophic level

1 Producers

2 Secondary consumers eat primary consumers/herbivores; secondary consumers are food/prey for tertiary consumers.

3 Decomposers break down the bodies of dead organisms; by secreting enzymes onto them; and absorbing the food molecules back inside themselves.

Pyramids of biomass

1 Pyramid should be drawn with a pencil and a ruler to scale e.g. oak tree (12 cm), caterpillars (5 cm) and blackbirds (2 cm).

2 $= \frac{400}{1200} \times 100\% \quad = 33.3\%$

3. Carbon dioxide and water are lost from the body as waste during exhalation; Water and urea are lost from the body as urine; Some food is not digestible and is lost from the body as faeces/by egestion.

Food production

1 Any one of: increasing birth rate; changing diets in developed countries; new pests and pathogens; environmental changes that affect food production/famine; the cost of agricultural inputs; conflicts which affect the availability of water or food.

2 If cod are caught then they cannot breed and the number of cod will decrease; If cod are fished only when there are a large number of mature adults, then some breeding will occur and maintain the numbers of cod.

3 Import food from another country; Some people could migrate to another country where water and food is available; Plant drought-resistant crops.

Role of biotechnology

1 Any one of: drought-resistant; pest-resistant; herbicide-resistant.

2 Microorganisms need oxygen; to carry out aerobic respiration.

3 Golden rice is genetically modified to contain beta-carotene; Beta-carotene is needed to make vitamin A; People who eat golden rice are less likely to suffer from vitamin A deficiency.

Review it!

1 a A group of organisms of the same species living in the same area at the same time.

b A group of populations living and interacting with each other in the same area.

c Any two from: temperature; light intensity; moisture levels in the soil / air; pH of the soil; wind intensity / direction; carbon dioxide levels; oxygen levels of the water.

2 a Used in photosynthesis; to make sugars.

b Aerobic respiration; and combustion

c i Decomposers respire and produce carbon dioxide.

ii Any one from: Optimum temperature; Increased water availability; Increased oxygen availability.

iii $1 \div 240 = 0.004\ s^{-1}$

3 a i Global temperatures are increasing; due to the increased amounts of greenhouse gases / carbon dioxide / methane in the atmosphere.

ii Water availability is decreasing; as higher temperatures causes desertification of some areas.

iii Carbon dioxide in the air increases as photosynthesis decreases due to fewer trees / plants.

b Global weather patterns will change causing decreased availability of food and water; sea levels will rise, decreasing habitats; increased extinction of species as some species will not be able to adapt quickly to the changing climate.

4 a Any two from: Increasing birth rate; changing diets in developed countries; pests / pathogens that affect farming

b Limit animals' movement; control the temperature of the surroundings; and feed animals high proteins food.

5 a Plankton b Herring

c Pyramid should be drawn with a pencil and ruler to scale, e.g. plankton (7 cm), shrimp (3 cm), herring (2 cm) and shark (1.5 cm).

d The number of sharks would decrease as there would be fewer herring to eat

6 a i Total number of buttercups – 34

ii $34 \div 10 = 3.4$;

Area = 10m × 15m = $150m^2$;

Estimated number of buttercups = $3.4 \times 150 = 510$ buttercups

b $= \frac{25 \times 18}{10} = 45$